1175/
1

D1205491

Hyperbolic Manifolds
and
Holomorphic Mappings

PURE AND APPLIED MATHEMATICS
A Series of Monographs

COORDINATOR OF THE EDITORIAL BOARD
S. Kobayashi
UNIVERSITY OF CALIFORNIA AT BERKELEY

1. KENTARO YANO. Integral Formulas in Riemannian Geometry (1970)
2. S. KOBAYASHI. Hyperbolic Manifolds and Holomorphic Mappings (1970)

In Preparation :

L. NARICI, E. BECKENSTEIN, and G. BACHMAN. Functional Analysis and Valuation Theory

V. S. VLADIMIROV. Equations of Mathematical Physics (A. Jeffrey, editor; A. Littlewood, translator)

Hyperbolic Manifolds
and
Holomorphic Mappings

SHOSHICHI KOBAYASHI

DEPARTMENT OF MATHEMATICS
UNIVERSITY OF CALIFORNIA
BERKELEY, CALIFORNIA

MARCEL DEKKER, INC., New York   1970

QA 331
.K717

INDIANA
UNIVERSITY
LIBRARY

NORTHWEST

Copyright © 1970 by MARCEL DEKKER, Inc.

All Rights Reserved

No part of this work may be reproduced or utilized in any form or by any means, electronic or mechanical, including photocopying, microfilm, and recording, or by any information storage and retrieval system, without permission in writing from the publisher.

MARCEL DEKKER, Inc.
95 Madison Avenue, New York, New York 10016

LIBRARY OF CONGRESS CATALOG CARD NUMBER: 70-131390

ISBN NO. 0-8247-1380-X

Printed in the United States of America

Dedicated to

Professor S. S. Chern and Professor K. Yano

Preface

This book is a development of lectures delivered in Berkeley in the academic year 1968–69. Its object is to give a coherent account of intrinsic pseudodistances on complex manifolds and of their applications to holomorphic mappings.

The classical Schwarz–Pick lemma states that every holomorphic mapping from a unit disk into itself is distance-decreasing with respect to the Poincaré distance. In Chapter I, we prove Ahlfors' generalization to holomorphic mappings from a unit disk into a negatively pinched Riemann surface and present some of its applications in the geometric theory of functions. In Chapters II and III, various higher-dimensional generalizations of the Schwarz–Pick–Ahlfors lemma are proved. The *raison d'être* of the first three chapters is to provide interesting examples for the subsequent chapters. It is therefore possible for the reader to start from Chapter IV and go back to Chapters I, II, and III only when he must.

In Chapter IV, we introduce a certain pseudodistance on every complex manifold in an intrinsic manner. A complex manifold is said to be (completely) hyperbolic if this pseudodistance is a (complete) distance. The classical pseudodistance of Carathéodory and this new pseudodistance share two basic properties: (1) they agree with the Poincaré distance on the unit disk, and (2) every holomorphic mapping is distance-decreasing. Among the pseudodistances with these two properties, the Carathéodory pseudodistance is the smallest and the new one is the largest. These pseudodistances permit us to obtain many results on complex manifolds by a purely metric space–topological method. They enable us also to gain a geometric insight into function theoretic results. Elementary properties of these pseudodistances and of hyperbolic manifolds are given.

In Chapter V, we study holomorphic mappings of a complex manifold into a hyperbolic manifold. In Chapter VI, which is, to a large extent, based on M. H. Kwack's thesis, we give generalizations of the big Picard theorem to higher-dimensional manifolds. Although there is more than one way to interpret the big Picard theorem geometrically, we consider it as an extension theorem for holomorphic mappings.

To avoid technical complications associated with complex spaces, we consider only complex manifolds in Chapters IV, V, and VI. In Chapter VII, we indicate how some of the results in these three chapters could be generalized to complex spaces.

In Chapter VIII, the relationships between hyperbolic manifolds and minimal models are studied. The generalized big Picard theorems are essentially used here. To a large extent this chapter is based on J. Zumbrunn's thesis.

Closely following the constructions of the pseudodistances in Chapter IV, we define in Chapter IX two kinds of intermediate dimensional measures on a complex manifold in a intrinsic manner. These measures have been studied more thoroughly by D. Eisenman in his thesis. Our approach is perhaps a little more differential geometric.

At the end of Chapter IX, we list a few unsolved problems on hyperbolic manifolds.

In preparing my lectures on hyperbolic manifolds, I had numerous useful conversations with H. Wu. By solving some of the problems listed in the first draft of this book, P. Kiernan has helped me to make a number of improvements. It was through Professor Chern's papers on holomorphic mappings that I was led into this topic. I wish to express my thanks to these mathematicians.

Berkeley, January, 1970   SHOSHICHI KOBAYASHI

Contents

I

The Schwarz Lemma and Its Generalizations

1. *The Schwarz–Pick Lemma*

Let D be the open unit disk in the complex plane $\mathbf{C}$, i.e.,

$$D = \{z \in \mathbf{C}; \; |z| < 1\}.$$

Let $f : D \to D$ be a holomorphic mapping such that $f(0) = 0$. Then the classical Schwarz lemma states

$$|f(z)| \leqq |z| \qquad \text{for} \quad z \in D$$

and

$$|f'(0)| \leqq 1,$$

and the equality $|f'(0)| = 1$ or the equality $|f(z)| = |z|$ at a single point $z \neq 0$ implies

$$f(z) = \varepsilon z \qquad \text{with} \quad |\varepsilon| = 1.$$

Now we shall drop the assumption $f(0) = 0$. If $f : D \to D$ is an arbitrary holomorphic mapping, we fix an arbitrarily chosen point $z \in D$ and consider the automorphisms g and h of D defined by

$$g(\zeta) = \frac{\zeta + z}{1 + \bar{z}\zeta} \qquad \text{for} \quad \zeta \in D,$$

$$h(\zeta) = \frac{\zeta - f(z)}{1 - \overline{f(z)}\,\zeta} \qquad \text{for} \quad \zeta \in D.$$

1

Then the composed mapping $F = h \cdot f \cdot g$ is a holomorphic mapping of D into itself which sends 0 into itself. Since $F(0) = 0$ and $F'(0) = h'(f(z))$ $f'(z) \, g'(0)$, we obtain

$$F'(0) = \frac{1 - |z|^2}{1 - |f(z)|^2} f'(z).$$

Hence,

$$\frac{1 - |z|^2}{1 - |f(z)|^2} |f'(z)| \leq 1,$$

or

$$\frac{|f'(z)|}{1 - |f(z)|^2} \leq \frac{1}{1 - |z|^2} \qquad \text{for} \quad z \in D.$$

We may conclude the following:

THEOREM 1.1. *Let f be a holomorphic mapping of the unit disk D into itself. Then*

$$\frac{|df|}{1 - |f|^2} \leq \frac{|dz|}{1 - |z|^2} \qquad \text{for} \quad z \in D,$$

and the equality at a single point z implies that f is an automorphism of D.

This result, which was derived from the Schwarz lemma, is actually equivalent to the Schwarz lemma. In fact, if $f : D \to D$ is a holomorphic mapping such that $f(0) = 0$, then by setting $z = 0$ in the inequality above, we obtain

$$|f'(0)| \leq 1,$$

and if $|f'(0)| = 1$, then f is an automorphism of D. Moreover,

$$\int_0^{|f(\zeta)|} \frac{|df|}{1 - |f|^2} \leq \int_0^{|\zeta|} \frac{|dz|}{1 - |z|^2}.$$

Hence,

$$\log \frac{1 + |f(\zeta)|}{1 - |f(\zeta)|} \leq \log \frac{1 + |\zeta|}{1 - |\zeta|},$$

which implies

$$\frac{2}{1 - |f(\zeta)|} - 1 = \frac{1 + |f(\zeta)|}{1 - |f(\zeta)|} \leq \frac{1 + |\zeta|}{1 - |\zeta|} = \frac{2}{1 - |\zeta|} - 1.$$

It follows that $|f(\zeta)| \leq |\zeta|$. The equality $|f(\zeta)| = |\zeta|$ at a single point $\zeta \neq 0$ implies the equality

$$\frac{|f'(z)|}{1 - |f(z)|^2} = \frac{1}{1 - |z|^2}$$

for all z lying on the line segment joining 0 and ζ. By Theorem 1.1, f is an automorphism of D, and hence, $f(z) = \varepsilon z$ for some ε with $|\varepsilon| = 1$. This proves that Theorem 1.1 implies the classical Schwarz lemma.

We shall now consider Theorem 1.1 from a differential geometric viewpoint. If we consider the Kaehler metric $ds_D{}^2$ on D given by

$$ds_D{}^2 = \frac{dz\, d\bar{z}}{(1 - |z|^2)^2},$$

then the inequality in Theorem 1.1 may be written as follows:

$$f^*(ds_D{}^2) \leq ds_D{}^2.$$

The metric $ds_D{}^2$ is called the *Poincaré metric* or the *Poincaré–Bergman metric* of D. Now, Theorem 1.1 may be stated as follows:

THEOREM 1.2.   *Let D be the open unit disk in $\mathbf{C}$ with the Poincaré–Bergman metric $ds_D{}^2$. Then every holomorphic mapping $f : D \to D$ is distance-decreasing, i.e., satisfies*

$$f^*(ds_D{}^2) \leq ds_D{}^2,$$

and the equality at a single point of D implies that f is an automorphism of D.

If f is an automorphism of D, then the Schwarz lemma applied to both f and f^{-1} implies that f is an isometry.

We note that the Gaussian curvature of the metric $ds_D{}^2$ is equal to -4 everywhere. [In general, the Gaussian curvature of a metric $2h\, dz\, d\bar{z}$ is given by $-(1/h)(\partial^2 \log h/\partial z\, \partial\bar{z}).$]

2. A Generalization by Ahlfors

Let D_a be the open disk of radius a in $\mathbf{C}$,

$$D_a = \{z \in \mathbf{C};\ |z| < a\}.$$

Then the metric

$$ds_a{}^2 = \frac{4a^2\,dz\,d\bar{z}}{A(a^2 - |z|^2)^2} \qquad (A > 0)$$

on D_a has Gaussian curvature $-A$. The following theorem of Ahlfors [1] generalizes the Schwarz lemma.

THEOREM 2.1. *Let M be a one-dimensional Kaehler manifold with metric $ds_M{}^2$ whose Gaussian curvature is bounded above by a negative constant $-B$. Then every holomorphic map $f: D_a \to M$ satisfies*

$$f^*\,ds_M{}^2 \leqq \frac{A}{B}\,ds_a{}^2.$$

Proof. Let u be the nonnegative function on D_a defined by

$$f^*(ds_M{}^2) = u\,ds_a{}^2.$$

We want to prove that $u \leqq A/B$ on D_a. Although u may not attain its maximum in (the interior of) D_a in general, we shall show that we have only to consider the case where u attains its maximum in D_a. Let r be a positive number less than a. Let z_0 be an arbitrary point of D_a. Taking r sufficiently close to a, we may assume that z_0 is in D_r. We denote by $ds_r{}^2$ the metric on D_r obtained from $ds_a{}^2$ by replacing a by r but keeping the same constant A. From the explicit expression for $ds_a{}^2$, it is clear that $(ds_r{}^2)_{z_0} \to (ds_a{}^2)_{z_0}$ as $r \to a$. If we define a nonnegative function u_r on D_r by $f^*(ds_M{}^2) = u_r\,ds_r{}^2$, then $u_r(z_0) \to u(z_0)$ as $r \to a$. Hence it suffices to prove that $u_r \leqq A/B$ on D_r. If we write $f^*(ds_M{}^2) = h\,dz\,d\bar{z}$ on D_a, then h is bounded on $\bar{D}_r$ (the closure of D_r). On the other hand, the coefficient $4r^2/A(r^2 - |z|^2)^2$ of $ds_r{}^2$ approaches infinity at the boundary of D_r. Hence, the function u_r defined on D_r goes to zero at the boundary of D_r. In particular, u_r attaints its maximum in D_r. The problem is thus reduced to the case where u attains its maximum in D_a.

We shall now impose the additional assumption that u attains its maximum in D_a, say at $z_0 \in D_a$. If $u(z_0) = 0$, then $u \equiv 0$ and there is nothing to prove. Assume that $u(z_0) > 0$. Then the mapping $f: D_a \to M$ is nondegenerate in a neighborhood of z_0 so that f is a biholomorphic map from an open neighborhood U of z_0 onto the open set $f(U)$ of M. Identifying U with $f(U)$ by the map f, we use the coordinate system z of $D_a \subset \mathbf{C}$ also as a local coordinate system in $f(U)$. If we write $ds_M{}^2$

$= 2h \, dz \, d\bar{z}$ on $f(U)$, then $f^*(ds_M{}^2) = 2h \, dz \, d\bar{z}$ on U. If we write $ds_a{}^2 = 2g \, dz \, d\bar{z}$, then $u = h/g$. The Gaussian curvature k of the metric $ds_M{}^2 = 2h \, dz \, d\bar{z}$ is given by

$$k = -\frac{1}{h} \frac{\partial^2 \log h}{\partial z \, \partial \bar{z}}.$$

The Gaussian curvature $-A$ of the metric $ds_a{}^2 = 2g \, dz \, d\bar{z}$ is given by

$$-A = -\frac{1}{g} \frac{\partial^2 \log g}{\partial z \, \partial \bar{z}}.$$

Since $k \leq -B$ by our assumption, we have

$$\frac{\partial^2 \log u}{\partial z \, \partial \bar{z}} = \frac{\partial^2 \log h}{\partial z \, \partial \bar{z}} - \frac{\partial^2 \log g}{\partial z \, \partial \bar{z}} = -kh - Ag \geqq Bh - Ag.$$

Since $\log u$ attains its maximum at z_0, the left-hand side in the inequality above is nonpositive at z_0 and so is the right-hand side. Hence, $A/B \geqq h/g = u$ at z_0. Since u attains its maximum at z_0, it follows that $A/B \geqq u$ everywhere. QED.

3. *The Gaussian Plane Minus Two Points*

In view of Theorem 2.1 we are naturally interested in finding a one-dimensional Kaehler manifold whose Gaussian curvature is bounded above by a negative constant. As we shall see later (see § 4 of Chapter IV), the Gaussian plane $\mathbf{C}$ cannot carry such a metric. The metric

$$ds^2 = 2(1 + |z|^2) \, dz \, d\bar{z}$$

on $\mathbf{C}$ has curvature $k = -1/(1 + |z|^2)^3$, which is strictly negative everywhere but is not bounded above by any negative constant.

If a one-dimensional complex manifold M carries a Kaehler metric whose Gaussian curvature is bounded above by a negative constant, so does any covering manifold of M. If M is the Gaussian plane minus a point, say the origin, that is, $M = \mathbf{C} - \{0\}$, then the universal covering manifold of M is $\mathbf{C}$, the covering projection being given by

$$z \in \mathbf{C} \rightarrow e^{2\pi i z} \in \mathbf{C} - \{0\}.$$

This shows that $\mathbf{C} - \{0\}$ does not admit a Kaehler metric whose Gaussian curvature is bounded above by a negative constant.

Consider now the Gaussian plane minus two points, say $M = \mathbf{C}$ $- \{0, 1\}$. If we use the modular function $\lambda(z)$, we can show that M carries a complete Kaehler metric with negative constant curvature. Let H denote the upper half-plane in $\mathbf{C}$, i.e., $H = \{z = x + iy \in \mathbf{C}; \ y > 0\}$. Then the modular function λ gives a covering projection $\lambda : H \to \mathbf{C}$ $- \{0, 1\}$. If we digress a little, the group of deck-transformations is given by

$$z \in H \to \frac{az + b}{cz + d} \in H$$

with

$$\begin{pmatrix} a & b \\ c & d \end{pmatrix} \in SL(2; \mathbf{Z}), \qquad \begin{pmatrix} a & b \\ c & d \end{pmatrix} \equiv \begin{pmatrix} 1 & 0 \\ 0 & 1 \end{pmatrix} \qquad \text{mod 2.}$$

This group, known as the congruence subgroup mod 2, is a normal subgroup of index 6 in the modular group $SL(2; \mathbf{Z})$, and its fundamental domain is given by F in Fig. 1. The boundary of F and the imaginary axis are mapped into the real axis by λ. In Fig. 2, one sees roughly how

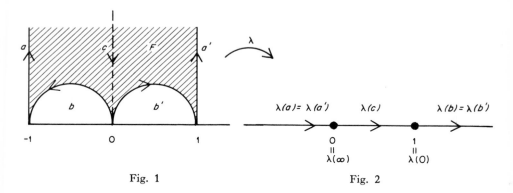

Fig. 1   Fig. 2

the fundamental domain F is mapped onto $\mathbf{C} - \{0, 1\}$ by λ. On the upper half-plane H, the metric $ds^2 = 2dz \, d\bar{z}/Ay^2$ has curvature $-A$ and is invariant by the group of holomorphic transformations of H. It follows immediately that this metric induces a metric of curvature $-A$ on $\mathbf{C} - \{0, 1\}$. Unfortunately, the modular function λ is so complicated that the induced metric on $\mathbf{C} - \{0, 1\}$ cannot be expressed in a simple form in terms of the natural coordinate of $\mathbf{C} - \{0, 1\}$. (For the definition and basic properties of the modular function λ, see for instance Ahlfors [2] and Ford [1].)

We shall now give a more elementary construction of a metric with Gaussian curvature ≤ -4 on $\mathbf{C} - \{0, 1\}$. The construction is due to Grauert and Reckziegel [1]. Given a positive C^∞ function $g(z, \bar{z})$ defined in an open set in $\mathbf{C}$, we define a real-valued function $K(g)$ defined in the same open set as follows:

$$K(g) = -\frac{1}{g} \frac{\partial^2 \log g}{\partial z\, \partial \bar{z}}.$$

The definition is motivated by the fact that $K(g)$ is the Gaussian curvature of the metric $ds^2 = 2g\, dz\, d\bar{z}$. We first prove the following:

PROPOSITION 3.1. *For positive functions f and g, we have*

(a) $c\, K(cg) = K(g)$ *for all positive numbers c;*
(b) $fg\, K(fg) = f\, K(f) + g\, K(g)$;
(c) $(f + g)^2\, K(f + g) \leq f^2\, K(f) + g^2\, K(g)$;
(d) *If* $K(f) \leq -k_1 < 0$ *and* $K(g) \leq -k_2 < 0$,

then

$$K(f + g) \leq -k_1 k_2/(k_1 + k_2).$$

Proof. (a) and (b) are immediate from the definition of $K(g)$. (c) follows from the following, which can be easily verified from the definition of $K(g)$.

$$fg(f + g)[f^2\, K(f) + g^2\, K(g) - (f + g)^2\, K(f + g)]$$
$$= \left| f \frac{\partial g}{\partial z} - g \frac{\partial f}{\partial z} \right|^2 \geq 0.$$

(d) is a consequence of (c) and

$$[f^2\, K(f) + g^2\, K(g)]/(f + g)^2 \leq -k_1 k_2/(k_1 + k_2).$$

This latter inequality follows from

$$-(f^2 k_1 + g^2 k_2)/(f + g)^2 \leq -k_1 k_2/(k_1 + k_2).$$

QED.

We set

$$p(z, \bar{z}) = |\, 2z\,|^{2a-2}(1 + |\, 2z\,|^{2a}),$$

where a is a constant, $0 < a \leq \frac{1}{5}$. Then p is a positive C^∞ function on $\mathbf{C} - \{0\}$. We have

$$K(p) = -4a^2/(1 + |\, 2z\,|^{2a})^3.$$

We set

$$f(z, \bar{z}) = p(z, \bar{z})\, p(z - 1, \bar{z} - 1).$$

Then f is a positive C^∞ function on $\mathbf{C} - \{0, 1\}$. By (b) of Proposition 3.1, we have

$$K[f(z, \bar{z})] = \frac{K[p(z, \bar{z})]}{p(z - 1, \bar{z} - 1)} + \frac{K[p(z - 1, \bar{z} - 1)]}{p(z, \bar{z})}.$$

Since

$$\frac{K[p(z, \bar{z})]}{p(z - 1, \bar{z} - 1)} \leq -\frac{a^2}{2(1 + 3^{2a})} \qquad \text{for} \quad 0 < |\, z\,| \leq \tfrac{1}{2}$$

and

$$\frac{K[p(z - 1, \bar{z} - 1)]}{p(z, \bar{z})} \leq -\frac{a^2}{2(1 + 3^{2a})} \qquad \text{for} \quad 0 < |\, z - 1\,| \leq \tfrac{1}{2},$$

we obtain

$$K[f(z, \bar{z})] \leq -\frac{a^2}{2(1 + 3^{2a})} \quad \text{if} \;\; 0 < |\, z\,| \leq \tfrac{1}{2} \;\; \text{or} \;\; 0 < |\, z - 1\,| \leq \tfrac{1}{2}.$$

On the other hand, for $|\, z - 1\,| \geq \tfrac{1}{2}$, we have

$$
\begin{aligned}
|\, 2z - 2\,|^{2a-2} & (1 + |\, 2z - 2\,|^{2a})(1 + |\, 2z\,|^{2a})^3 \\
&= |\, 2z - 2\,|^{2a-2/5}(|\, 2z - 2\,|^{-2/5} + |\, 2z - 2\,|^{2a-2/5}) \\
&\quad \times (|\, 2z - 2\,|^{-2/5} + |\, 2z\,|^{2a}\,|\, 2z - 2\,|^{-2/5})^3 \\
&\leq 1 \cdot (1 + 1)(1 + |\, 2z\,|^{2a}\,|\, 2z - 2\,|^{-2a})^3 \\
&\leq 2(1 + 3^{2a})^3.
\end{aligned}
$$

Similarly, for $|\, z\,| \geq \tfrac{1}{2}$, we have

$$|\, 2z\,|^{2a-2}(1 + |\, 2z\,|^{2a})(1 + |\, 2z - 2\,|^{2a})^3 \leq 2(1 + 3^{2a})^3.$$

It follows that if $|z| \geq \frac{1}{2}$ and $|z - 1| \geq \frac{1}{2}$, then

$$
\begin{aligned}
K[f(z, \bar{z})] &= \frac{K[p(z, \bar{z})]}{p(z - 1, \bar{z} - 1)} + \frac{K[p(z - 1, \bar{z} - 1)]}{p(z, \bar{z})} \\
&\leq -\frac{2a^2}{(1 + 3^{2a})^3} - \frac{2a^2}{(1 + 3^{2a})^3} \\
&= -\frac{4a^2}{(1 + 3^{2a})^3} .
\end{aligned}
$$

Since $-4a^2/(1 + 3^{2a})^3 \leq -a^2/2(1 + 3^{2a})$ for $0 < a \leq \frac{1}{5}$, we have finally

$$
K[f(z, \bar{z})] \leq -\frac{a^2}{2(1 + 3^{2a})} \qquad \text{for} \quad z \in \mathbf{C} - \{0, 1\}.
$$

We have shown that the metric $ds^2 = 2f(z, \bar{z}) \, dz \, d\bar{z}$ on $\mathbf{C} - \{0, 1\}$ has Gaussian curvature $K(f) \leq -a^2/2(1 + 3^{2a}) < 0$. But this metric is not complete. Since $ds = \sqrt{2f} |dz|$ has a singularity of order $|z|^{a-1}$ at 0, the point 0 is at a finite distance from any point of $\mathbf{C} - \{0, 1\}$. Similarly, for the point 1. It is also easy to verify that the point at infinity of $\mathbf{C}$ is at a finite distance from any point of $\mathbf{C} - \{0, 1\}$ with respect to the metric ds^2. We shall make an adjustment to obtain a complete metric whose Gaussian curvature is bounded above by a negative constant.

Choose a real-valued C^∞ function $s(z, \bar{z})$ on $\mathbf{C}$ such that

$$
\begin{aligned}
s(z, \bar{z}) &= 1 && \text{for} \quad |z| \leq \tfrac{1}{4}, \\
0 \leq s(z, \bar{z}) &\leq 1 && \text{for} \quad \tfrac{1}{4} \leq |z| \leq \tfrac{1}{3}, \\
s(z, \bar{z}) &= 0 && \text{for} \quad |z| \geq \tfrac{1}{3}.
\end{aligned}
$$

We set

$$
\begin{aligned}
q(z, \bar{z}) &= s(z, \bar{z})/|z|^2 (\log |z|^2)^2 && \text{for} \quad 0 < |z| \leq \tfrac{1}{3}, \\
&= 0 && \text{for} \quad |z| \geq \tfrac{1}{3}.
\end{aligned}
$$

We set

$$
h(z, \bar{z}) = q(z, \bar{z}) + q(z - 1, \bar{z} - 1) + \frac{1}{|z|^4} q\left(\frac{1}{z}, \frac{1}{\bar{z}}\right).
$$

Since $K[q(z, \bar{z})] = -2$ for $0 < |z| < \tfrac{1}{4}$, we obtain

$$
K[h(z, \bar{z})] = -2 \text{ on } \{0 < |z| < \tfrac{1}{4}\} \cup \{0 < |z-1| < \tfrac{1}{4}\} \cup \{|z| > 4\}.
$$

We set

$$g = f + ch, \qquad 0 < c < 1.$$

If we denote $a^2/2(1 + 3^{2a})$ by k' so that $K(f) \leq -k'$ on $\mathbf{C} - \{0, 1\}$, then by (d) of Proposition 3.1 we have

$$K[g(z, \bar{z})] \leq -\frac{2k'}{2 + ck'} < -\frac{2k'}{2 + k'} < 0$$

on the domain $\{0 < |z| < \frac{1}{4}\} \cup \{0 < |z - 1| < \frac{1}{4}\} \cup \{|z| > 4\}$. In the complementary set which is compact, we use the following estimate [cf. (c) of Proposition 3.1]:

$$
\begin{aligned}
K(g) &= K[c(f + h) + (1 - c)f] \\
&\leq \frac{c^2(f + h)^2 K[c(f + h)] + (1 - c)^2 f^2 K[(1 - c)f]}{(f + ch)^2} \\
&= \frac{c(f + h)^2 K(f + h) + (1 - c)f^2 K(f)}{(f + ch)^2} \\
&\leq \frac{c(f + h)^2 K(f + h) - (1 - c)f^2 k'}{(f + ch)^2}.
\end{aligned}
$$

If we take a sufficiently small c, then $K(g)$ is bounded above by a negative constant on the compact set we are considering.

We have thus constructed a complete metric $ds^2 = 2g \, dz \, d\bar{z}$ on $\mathbf{C} - \{0, 1\}$ whose Gaussian curvature $K(g)$ is bounded above by a negative constant, say k. Multiplying the metric by a suitable constant, we may assume that $k = -4$.

4. Schottky's Theorem

One of the best known consequences of the generalized Schwarz lemma is the following theorem of Schottky [1].

THEOREM 4.1. *Given a complex number a with $a \neq 0, 1$ and a real number r with $0 \leq r < 1$, there exists a positive number $S = S(a, r)$ with the following property: If f is a holomorphic map from the open unit disk $D = \{z \in \mathbf{C}; |z| < 1\}$ into $\mathbf{C} - \{0, 1\}$ such that $f(0) = a$, then $|f(z)| \leq S(a, r)$ for $|z| \leq r$.*

Proof. Let ds_D^2 denote the Poincaré metric on D with curvature -4. We set $M = \mathbf{C} - \{0, 1\}$ and let ds_M^2 be a complete metric with curvature ≤ -4; such a metric was constructed in the preceding section. If we set $r' = \frac{1}{2} \log(1+r)/(1-r)$, then the set $\{z \in \mathbf{C}; |z| < r\}$ coincides with the set of points in D whose distances from 0 with respect to ds_D^2 do not exceed r'. Let $N(a; r')$ be the set of points in M which are at distance r' or less from a with respect to the metric ds_M^2. Since the metric ds_M^2 is complete, $N(a; r')$ is compact and hence is contained in the set $\{z \in \mathbf{C}; |z| \leq S\}$ for a suitable positive number S. Let $f : D \to M$ be a holomorphic map such that $f(0) = a$. Since f is distance-decreasing, i.e., $f^*(ds_M^2) \leq ds_D^2$ by Theorem 2.1, it maps the set $\{z \in D; |z| \leq r\}$ into $N(a, r')$.   QED.

Since the construction of the metric ds_M^2 in the preceding section is fairly explicit, we can give an estimate on $S(a, r)$. Let $ds_M^2 = 2g\,dz\,d\bar{z}$ be the complete metric with curvature ≤ -4 constructed on $\mathbf{C} - \{0, 1\}$ in the preceding section. From our construction, we see that $g > A^2h$ on $\mathbf{C} - \{0, 1\}$ for a suitable positive number A, where h was defined in § 3. From the definition of h, we see that, in the domain $\{z \in \mathbf{C}; |z| \geq 3\}$, h coincides with

$$\frac{1}{|z|^2(\log|z|^2)^2}.$$

If we use the polar coordinate system (r, θ), then we have

$$ds_M^2 = 2g\,dz\,d\bar{z} > 2A^2h\,dz\,d\bar{z}$$
$$\geq \frac{A^2\,dr\,dr}{2r^2(\log r)^2} \qquad \text{on} \quad \{z \in \mathbf{C}; |z| \geq 3\}.$$

Let a and b be two points of $\mathbf{C}$ such that $3 \leq |a| \leq |b|$. Then the distance from a to b with respect to ds_M^2 is greater than

$$\int_{|a|}^{|b|} \frac{A\,dr}{\sqrt{2}\,r\log r} = \frac{A}{\sqrt{2}} \log \frac{\log|b|}{\log|a|}.$$

If $|a| < 3$, then it suffices to replace $\log|a|$ by $\log 3$. On the other hand, the distance from $0 \in D$ to $z \in D$ with respect to ds_D^2 is equal to

$$\frac{1}{2} \log \frac{1+|z|}{1-|z|}.$$

Let $f : D \to M = \mathbf{C} - \{0, 1\}$ be a holomorphic map. Since it is distance-decreasing, we have

$$\frac{A}{\sqrt{2}} \log \frac{\log |f(z)|}{\log |f(0)|} < \frac{1}{2} \log \frac{1+|z|}{1-|z|},$$

provided that $|f(0)| \geq 3$ and $|f(z)| \geq 3$. If $|f(0)| < 3$ and $|f(z)| \geq 3$, it suffices to replace $\log |f(0)|$ by $\log 3$. In general, we obtain

$$\log |f(z)| < \mathrm{Max}\left\{\log 3, \ B\left(\frac{1+|z|}{1-|z|}\right)^{C}\right\},$$

where $B = \mathrm{Max}\{\log 3, \log|f(0)|\}$ and $C = \sqrt{2}/A$.

If we wish to estimate the constant C, we have to write down the function $s(z, \bar{z})$ in § 3 explicitly.

Similar estimates on $\log f(z)$ have been obtained by Ostrowski [1], Pfluger [1], and Ahlfors [1]. See also Heins [1].

5. *Compact Riemann Surfaces of Genus* ≤ 2

In this section we prove the following statement:

THEOREM 5.1. *Every compact Riemann surface M of genus $g \geq 2$ admits a Kaehler metric whose Gaussian curvature is bounded above by a negative constant.*

Proof. If we make use of the classical result that every compact Riemann surface M of genus $g \geq 2$ is a quotient of the upper half-plane H by a discontinuous group Γ of linear fractional transformations acting freely on H, we see easily that M admits a Kaehler metric of constant negative curvature. We shall give here a more elementary proof following Grauert and Reckziegel [1].

Since the genus of M is g, there are g linearly independent holomorphic 1-forms $\omega_1, \ldots, \omega_g$ on M. We set

$$ds^2 = \omega_1 \bar{\omega}_1 + \omega_2 \bar{\omega}_2.$$

In terms of a local coordinate system z of M, we may write

$$ds^2 = (f_1 \bar{f}_1 + f_2 \bar{f}_2)\, dz\, d\bar{z}, \qquad \text{where} \qquad \omega_i = f_i\, dz.$$

If $p_1, \ldots, p_k$ is the set of common zeros of ω_1 and ω_2, then ds^2 is a

Kaehler metric on $M - \{p_1, \ldots, p_k\}$. By a simple calculation we see that the Gaussian curvature K is given by

$$K = -2 \,|\, f_1'f_2 - f_1 f_2' \,|^2 / (f_1 \bar{f}_1 + f_2 \bar{f}_2)^3.$$

Since ω_1 and ω_2 are linearly independent, it follows that $f_2 \neq 0$ and $f_1/f_2 \neq$ constant, and consequently $K \not\equiv 0$. Hence there are at most finitely many points $p_{k+1}, \ldots, p_m$ in $M - \{p_1, \ldots, p_k\}$ where the curvature K vanishes. For each p_i, choose a neighborhood V_i in such a way that $V_i \cap V_j = \emptyset$ for $i \neq j$. In each V_i we shall modify the metric so that the Gaussian curvature becomes negative everywhere on M. We fix i, $i = 1, \ldots, m$. Choosing a local coordinate system z in V_i such that $z(p_i) = 0$, let r be a positive number and

$$B = \{z; \,|\, z \,|\, < r\} \Subset V_i, \qquad B' = \{z; \,|\, z \,|\, < r/2\}.$$

Let $a(z, \bar{z})$ be a C^∞ function on V_i such that

$$0 \leq a(z, \bar{z}) \leq 1, \qquad a \,|\, B' \equiv 1, \qquad a \,|\, (V_i - B) \equiv 0.$$

Let c be a constant, $0 < c < 1$, and set

$$g = f + ch, \quad \text{where} \quad f = f_1 \bar{f}_1 + f_2 \bar{f}_2 \quad \text{and} \quad h = a \cdot (1 + z\bar{z}).$$

Then the metric $g \, dz \, d\bar{z}$ coincides with ds^2 on $V_i - B$. Since the curvature of ds^2 is bounded above by a negative constant on the compact set $\bar{B} - B'$, the metric $g \, dz \, d\bar{z}$ has a strictly negative curvature on $\bar{B} - B'$ if c is sufficiently small. From the estimate on $K(g)$ given at the end of Section 3, we see that the curvature of $g \, dz \, d\bar{z}$ is strictly negative in B' if c is sufficiently small.   QED.

Remark. It is known (see, for instance, Springer [1, p. 270]) that if the genus $g \geq 1$, there is no point of M where all holomorphic 1-forms vanish. It is therefore possible to choose ω_1 and ω_2 in the foregoing proof in such a way that $\omega_1 \bar{\omega}_1 + \omega_2 \bar{\omega}_2$ is positive definite everywhere on M. But this does not simplify the proof.

6. *Holomorphic Mappings from an Annulus into an Annulus*

Let A be the annulus in $\mathbf{C}$ defined by

$$A = \{z \in \mathbf{C}; \, 0 < r < |\, z \,|\, < R\}.$$

The number $M = \log(R/r)$ is called the *modulus* of A. Two annuli with the same modulus are clearly biholomorphic to each other under a suitable homothetic transformation. We may therefore assume that $rR = 1$.

Let A_1 and A_2 be two annuli with moduli M_1 and M_2, respectively:

$$A_k = \{z \in \mathbf{C};\ 0 < r_k < |z| < 1/r_k\}, \qquad k = 1, 2.$$

Let $f : A_1 \to A_2$ be a holomorphic mapping and $f_* : \pi_1(A_1) \to \pi_1(A_2)$ the induced homomorphism on the fundamental groups of A_1 and A_2. Let α_k be the generator of $\pi_1(A_k) = Z$, $k = 1, 2$. Then the *degree of f*, denoted by $\deg f$, is defined by

$$f_*(\alpha_1) = (\deg f)\alpha_2.$$

For each integer m such that $|m| \leq M_2/M_1$, we define a holomorphic mapping $f_m : A_1 \to A_2$ with $\deg f = m$ as follows:

$$f_m(z) = z^m \qquad \text{for} \quad z \in A_1.$$

From topology we know that any mapping $f : A_1 \to A_2$ (holomorphic or not) of degree m with $|m| \leq M_2/M_1$ is homotopic to f_m. In fact, any two mappings of A_1 into A_2 with the same degree are homotopic to each other. We know also that for any integer m there is a mapping (not necessarily holomorphic) of A_1 into A_2 with degree m. But we have

THEOREM 6.1. *Let A_1 and A_2 be annuli with moduli M_1 and M_2 as above. If $f : A_1 \to A_2$ is a holomorphic mapping, then $|\deg f| \leq M_2/M_1$. If M_2/M_1 is a integer, then a holomorphic mapping $f : A_1 \to A_2$ with degree $m = \pm M_2/M_1$ coincides with the mapping f_m up to a rotation.*

Proof. We consider the band $B = \{z \in \mathbf{C};\ -b < \operatorname{Im} z < b\}$ of width $2b$. We know that it is conformally equivalent to the open unit disk or equivalently to the upper half-plane $H = \{w \in \mathbf{C};\ \operatorname{Im} w > 0\}$ in $\mathbf{C}$. Indeed, the mapping $z \in B \to ie^{\pi z/2b} \in H$ is a biholomorphic mapping. The invariant metric ds_H^2 of curvature -1 on H is given by

$$ds_H^2 = \frac{dw\, d\bar{w}}{v^2} \qquad \text{where} \quad w = u + iv.$$

If we induce this metric to the band B by the holomorphic mapping given above, we obtain the following invariant metric $ds_B{}^2$ of curvature -1:

$$ds_B{}^2 = \frac{\pi^2 \, dz \, d\bar{z}}{4b^2 \cos^2(\pi y/b)} \qquad \text{where} \quad z = x + iy.$$

We consider now a holomorphic mapping p from B onto the annulus $A = \{w \in \mathbf{C}; \, r < |\, w \,| < 1/r\}$, $r = e^{-2\pi b}$, defined by

$$p(z) = e^{2\pi i z} \qquad z \in B.$$

Then $p : B \to A$ is a covering projection. We denote by $ds_A{}^2$ the metric on A induced by $ds_B{}^2$. Then the rectangle $\{z \in B; \, 0 \leq x \leq 1\}$ is a fundamental domain for this covering space (B, A, p). The projection p maps the upper edge, the lower edge, and the two vertical edges of this rectangle onto the inner boundary, the outer boundary of the annulus A, and the segment $\{w = u + iv \in A; \, u > 0, \, v = 0\}$, respectively. It is also clear that p maps $\{z = x + iy \in B; \, 0 \leq x \leq 1, \, y = 0\}$ onto the unit circle $\{w \in A; \, |\, w \,| = 1\}$, which is the generator of the fundamental group $\pi_1(A)$. Consider a curve $w(t)$, $0 \leq t \leq 1$, in A which represents the generator of $\pi_1(A)$. We may assume that $w(0) = w(1) > 0$. To compute its length with respect to $ds_A{}^2$, we consider a curve $z(t)$ in the rectangle $\{z \in B; \, 0 \leq x \leq 1\}$ such that $p[z(t)] = w(t)$, $\mathrm{Re}[z(0)] = 0$ and $\mathrm{Re}[z(1)] = 1$, and compute the length of $z(t)$ with respect to $ds_B{}^2$. From the expression of $ds_B{}^2$ given above, we see that the curve $w(t)$ has the least length when it is the circle $|\, w(t) \,| = 1$, i.e., when $z(t)$ is real for all t. This least length is given by

$$\int_0^1 \frac{\pi \, dx}{2b} = \frac{\pi}{2b}.$$

We have shown that the circle $w(t) = e^{2\pi i t}$, $0 \leq t \leq 1$, represents the generator α of $\pi_1(A)$ and has arc-length $\pi/2b$ with respect to $ds_A{}^2$ and that any closed curve in A representing α has arc-length $\geq \pi/2b$, where the equality holds only when the closed curve coincides with the unit circle up to a parametrization. Similarly, the curve $w(t) = e^{2m\pi i t}$, $0 \leq t \leq 1$, representing $m\alpha \in \pi_1(A)$ has arc-length $|\, m\pi/2b \,|$ and any closed curve representing $m\alpha$ has arc-length $> |\, m\pi/2b \,|$ unless it coincides with $w(t) = e^{2m\pi i t}$ up to a parametrization.

Given two annuli A_1 and A_2, we consider the bands B_1 and B_2 of widths b_1 and b_2 and covering projections $p_1 : B_1 \to A_1$ and $p_2 : B_2 \to A_2$

as above. Let $f : A_1 \to A_2$ be a holomorphic mapping. Since B_1 is holomorphically equivalent to the unit open disk, we may apply Schwarz–Pick–Ahlfors lemma (Theorem 2.1) to the map $f \circ p_1 : B_1 \to A_2$ and see that $f \circ p_1$ is distance-decreasing, i.e., $(f \circ p_1)^* \, ds^2_{A_2} \leq ds^2_{B_1}$. Since p_1 is a local isometry, we see that f itself is also distance-decreasing, i.e., $f^* \, ds^2_{A_2} \leq ds^2_{A_1}$. We consider the unit circle $w_1(t) = e^{2\pi i t}$, $0 \leq t \leq 1$, in A_1 and its image curve $w_2(t) = f[w_1(t)]$, $0 \leq t \leq 1$, in A_2. Since $w_1(t)$ has arc-length $\pi/2b_1$ and f is distance-decreasing, $w_2(t)$ has arc-length $\leq \pi/2b_1$. If $\deg f = m$ so that $w_2(t)$ represents $m\alpha_2 \in \pi_1(A_2)$, where α_2 is the generator of $\pi_1(A_2)$, then $w_2(t)$ has arc-length $\geq |m\pi/2b_2|$. Hence, $|m\pi/2b_2| \leq \pi/2b_1$, that is, $|m| \leq b_2/b_1$. Since $r_k = e^{-2\pi b_k}$ and $M_k = \log(1/r_k{}^2) = -2 \log r_k$ for $k = 1$, 2, we have $b_2/b_1 = M_2/M_1$. Hence, $|m| \leq M_2/M_1$, which proves the first assertion of the theorem.

Assume that M_2/M_1 is an integer and $|m| = M_2/M_1$. Then $w_2(t)$ has arc-length $|m\pi/2b_2| = \pi/2b_1$. Hence, the closed curve $w_2(t)$ coincides with $e^{2m\pi i t}$, $0 \leq t \leq 1$, up to a parametrization. Since f is distance-decreasing and $w_2(t)$ has arc-length $\pi/2b_1$, f maps $w_1(t)$ onto $w_2(t)$ isometrically. This implies that $w_2(t)$ coincides actually with $e^{2m\pi i t}$ up to a rotation. It follows that f and f_m coincide on the unit circle $w_1(t)$ of A_1 up to a rotation and hence they coincide on A_1 up to a rotation.   QED.

COROLLARY 6.2. *Let* f *be a holomorphic mapping from a annulus* $A = \{z \in \mathbf{C}; \, r < |z| < 1/r\}$ *into itself. Then either* f *is homotopic to a constant map or is of the form* $f(z) = e^{\pm 2\pi i(z+a)}$, *where* a *is a real number.*

Theorem 6.1 is due to Huber [1]. The proof presented here is perhaps a little more differential geometric. In connection with the result of this section, see also Schiffer [1], Jenkins [1], and Landau and Osserman [1, 2].

II

Volume Elements and the Schwarz Lemma

1. Volume Element and Associated Hermitian Form

Let M be a complex manifold of complex dimension n and v_M a volume element of M. By definition v_M is a $2n$-form which is positive in the following sense. In terms of a local coordinate system $z^1, \ldots, z^n$ of M we may write

$$v_M = i^n K \, dz^1 \wedge d\bar{z}^1 \wedge \cdots \wedge dz^n \wedge d\bar{z}^n,$$

where K is a positive function. To this volume element v_M we associate the 2-form φ_M defined by

$$\varphi_M = 2i \sum R_{\alpha\bar{\beta}} \, dz^\alpha \wedge d\bar{z}^\beta,$$

where

$$R_{\alpha\bar{\beta}} = -\partial^2 \log K / \partial z^\alpha \, \partial \bar{z}^\beta,$$

and also the Hermitian form h_M defined by

$$h_M = 2 \sum R_{\alpha\bar{\beta}} \, dz^\alpha \, d\bar{z}^\beta.$$

For the reason given below, we shall call h_M the *Ricci tensor* of M (with respect to the volume element v_M).

Example 1.   Let M be an n-dimensional Kaehler manifold with metric $ds_M^2 = \sum 2g_{\alpha\bar{\beta}} \, dz^\alpha \, d\bar{z}^\beta$. The volume element v_M of M is given by

$$v_M = i^n G \, dz^1 \wedge d\bar{z}^1 \wedge \cdots \wedge dz^n \wedge d\bar{z}^n,$$

17

where
$$G = \det(g_{\alpha\bar\beta}).$$

Then the Ricci tensor (or rather its components) is given by (see, for instance, Kobayashi and Nomizu [1, Vol. II])

$$R_{\alpha\bar\beta} = -\partial^2 \log G / \partial z^\alpha \, \partial \bar z^\beta.$$

What we have said so far is true also for a Hermitian manifold, provided that we use the Hermitian connection. (For details, see § 1 of Chapter III).

Example 2. Let M be an n-dimensional complex manifold and H the Hilbert space of holomorphic n-forms φ such that

$$\int_M i^{n^2}\varphi \wedge \bar\varphi < \infty.$$

The inner product in H is defined by

$$(\varphi, \psi) = \int_M i^{n^2}\varphi \wedge \bar\psi \qquad \text{for} \quad \varphi, \psi \in H.$$

Let $\varphi_0, \varphi_1, \varphi_2, \ldots$ be an orthonormal basis for H (provided that H is nontrivial). The $2n$-form v_M defined by

$$v_M = i^{n^2} \sum_j \varphi_j \wedge \bar\varphi_j$$

is independent of the choice of an orthonormal basis. If for every point x of M there is an element $\varphi \in H$ which does not vanish at x, then v_M is a volume element of M and is called the *Bergman kernel form*. The Hermitian form h_M associated with v_M is easily seen to be negative semidefinite. If h_M is negative definite, $-h_M$ is called the *Bergman metric* of M. Since the Bergman metric $-h_M$ is a Kaehler metric, it gives rise to a volume element in the usual sense, which will be denoted by v'_M. We shall denote by h'_M the Hermitian form associated with v'_M. From these constructions we see that v_M, h_M, v'_M, and h'_M are all invariant by the group of holomorphic transformations of M. If M is homogeneous, i.e., if the group of holomorphic transformations is transitive on M, then $v'_M = cv_M$, where c is a constant. It follows that if M is homogeneous, then $h'_M = h_M$. In other words, if $g_{\alpha\bar\beta}$ and $R_{\alpha\bar\beta}$ denote the components of the Bergman metric and of its Ricci tensor, the homogeneity of M implies $R_{\alpha\bar\beta} = -g_{\alpha\bar\beta}$. If M is a bounded domain in $\mathbf{C}^n$ with Euclidean coordinate system $z^1, \ldots, z^n$, then H may be identified in a natural manner with the space of square integrable holomorphic functions on M with respect to the Lebesgue measure $i^n\, dz^1 \wedge d\bar z^1 \wedge \cdots \wedge dz^n \wedge d\bar z^n$. For a bounded domain M of $\mathbf{C}^n$, v_M is always positive and $-h_M$ is always positive-definite. If we write $v_M = i^n\, K(z, \bar z)\, dz^1 \wedge d\bar z^1 \wedge \cdots \wedge dz^n \wedge d\bar z^n$, we call $K(z, \bar z)$ the *Bergman kernel function* of the domain M.

For more details, see Kobayashi [1] and Lichnerowicz [1] as well as Bergman [1]. For homogeneous M, see Koszul [1], Hano and Kobayashi [1], and Pyatetzki-Shapiro [1].

The closed 2-form $(1/4\pi)\varphi_M$ is known to represent the first Chern class $c_1(M)$ of M (see Chern [2] and Kobayashi and Nomizu [1, Vol. 2]).

2. *Basic Formula*

Let M and M' be two complex manifolds of dimension n with volume elements v_M and $v_{M'}$, respectively. Let h_M and $h_{M'}$ be the Ricci tensors associated to v_M and $v_{M'}$, respectively.

Let $f : M' \to M$ be a holomorphic mapping and define the ratio $u = f^*(v_M)/v_{M'}$ by $f^*(v_M) = u \cdot v_{M'}$. Then u is a nonnegative function on M'. At a point where u is positive, i.e., a point where f is locally biholomorphic, we want to calculate the complex Hessian of $\log u$. Let $p \in M'$ be such a point and choose a local coordinate system $z^1, \cdots, z^n$ in a neighborhood of p and a local coordinate system $w^1, \ldots, w^n$ in a neighborhood of $f(p)$ in such a way that the map f is given by $w^i = z^i$, $i = 1, \ldots, n$. We write

$$v_M = i^n K\, dw^1 \wedge d\bar w^1 \wedge \cdots \wedge dw^n \wedge d\bar w^n,$$
$$v_{M'} = i^n K'\, dz^1 \wedge d\bar z^1 \wedge \cdots \wedge dz^n \wedge d\bar z^n.$$

Then

$$u = f^*K/K'$$

and

$$2 \sum_{\alpha,\beta} \frac{\partial^2 \log u}{\partial z^\alpha\, \partial \bar z^\beta}\, dz^\alpha\, d\bar z^\beta = h_{M'} - f^* h_M.$$

We have proved (Kobayashi [2]).

THEOREM 2.1. *Let M and M' be complex manifolds of dimension n with volume elements v_M and $v_{M'}$ and Ricci tensors h_M and $h_{M'}$, respectively. Let $f : M' \to M$ be a holomorphic mapping and let $u = f^*(v_M)/v_{M'}$. Then at a point of M' where u is different from zero, the following formula holds:*

$$2 \sum_{\alpha,\beta} \frac{\partial^2 \log u}{\partial z^\alpha \, \partial \bar{z}^\beta} \, dz^\alpha \, d\bar{z}^\beta = h_{M'} - f^* h_M ,$$

where $z^1, \ldots, z^n$ is a local coordinate system at the point.

In general, for a Hermitian form h on a complex vector space V we denote by $\dim^+ h$ (respectively, $\dim^- h$) the dimension of a maximal dimensional complex subspace of V on which h is positive (negative) definite. Hence h is positive (negative) definite on V if and only if $\dim^+ h = \dim V$ ($\dim^- h = \dim V$).

PROPOSITION 2.2. *Let $f : M' \to M$, h_M, $h_{M'}$, and u be as in Theorem 2.1. Then we have*

(1) *If $\dim^+ h_{M'} > \dim^+ h_M$ or $\dim^- h_{M'} < \dim^- h_M$ everywhere, the function u attains no nonzero local maximum on M';*

(2) *If $\dim^+ h_{M'} < \dim^+ h_M$ or $\dim^- h_{M'} > \dim^- h_M$ everywhere, the function u attains no nonzero local minimum on M'.*

Proof. Assume that u attains a nonzero local maximum at $p \in M'$. Then the Hermitian matrix $(\partial^2 \log u/\partial z^\alpha \, \partial \bar{z}^\beta)$ is negative semidefinite at p. By Theorem 2.1, $h_{M'} - f^* h_M$ is negative semidefinite at p. Since $u(p) \neq 0, f^*$ is nondegenerate at p so that $\dim^+ f^* h_M(p) = \dim^+ h_M[f(p)]$ and $\dim^- f^* h_M(p) = \dim^- h_M[f(p)]$. Hence,

$$\dim^+ h_{M'}(p) \leqq \dim^+ h_M[f(p)], \quad \dim^- h_{M'}(p) \geqq \dim^- h_M[f(p)].$$

This proves (1). The proof for (2) is similar.   QED.

Remark. If M' is a Hermitian manifold with Hermitian metric $ds^2 = 2 \sum g_{\alpha\bar{\beta}} \, dz^\alpha \, d\bar{z}^\beta$, then the complex Laplacian $\square(\log u)$ of $\log u$ is defined by

$$\square(\log u) = \sum g^{\alpha\bar{\beta}} \frac{\partial^2 \log u}{\partial z^\alpha \, \partial \bar{z}^\beta} .$$

From Theorem 2.1 we obtain

$$\square(\log u) = R_{M'} - \tfrac{1}{2} \operatorname{Trace}(f^* h_M),$$

where $R_{M'}$ is the scalar curvature of M' and $\mathrm{Trace}(f^*h_M)$ denotes the trace of f^*h_M with respect to the metric ds^2 of M'. This formula was proved by Chern [2]. If M' is a Kaehler manifold, then $\square(\log u) = \frac{1}{2} \triangle(\log u)$, where $\triangle$ is the ordinary Laplacian.

3. Holomorphic Mappings $f : M' \to M$ with Compact M'

We shall apply the results of § 2 to the case where M' is a compact complex manifold.

THEOREM 3.1. *Let M and M' be n-dimensional complex manifolds with volume elements v_M and $v_{M'}$ and Ricci tensor h_M and $h_{M'}$, respectively. Let $f : M' \to M$ be a holomorphic mapping. Assume that M' is compact. Then*

(1) *If $\dim^+ h_{M'} > \dim^+ h_M$ or $\dim^- h_{M'} < \dim^- h_M$ everywhere, the mapping f is everywhere degenerate, i.e., $f^*v_M = 0$;*

(2) *If $\dim^+ h_{M'} < \dim^+ h_M$ or $\dim^- h_{M'} > \dim^- h_M$ everywhere, the mapping f is degenerate at some point of M', i.e., f^*v_M vanishes at some point of M'.*

Proof. (1) Assuming the contrary, let p be a point of M' where $u = f^*v_M/v_{M'}$ attains its maximum. Then $u(p) \neq 0$. Now, our assertion follows from (1) of Proposition 2.2.

(2) Similarly, (2) follows from (2) of Proposition 2.2.   QED.

COROLLARY 3.2. *Let $f : M' \to M$ be as in Theorem 3.1. Then*

(1) *If the Ricci tensor $h_{M'}$ is everywhere positive definite on M' and if the Ricci tensor h_M is nowhere positive definite on M, the mapping f is degenerate everywhere on M';*

(2) *If h_M is everywhere negative definite on M and if $h_{M'}$ is nowhere negative definite on M', f is degenerate everywhere on M';*

(3) *If h_M is everywhere positive definite on M and if $h_{M'}$ is nowhere positive definite on M', f is degenerate somewhere on M';*

(4) *If $h_{M'}$ is everywhere negative definite on M' and if h_M is nowhere negative definite on M, f is degenerate somewhere on M'.*

Let φ_M be the 2-form associated to v_M (see § 1). Let φ_M^n be its nth exterior power and define a function r_M by

$$\frac{1}{2^n}\, \varphi_M^n = r_M v_M,$$

so that, locally,

$$r_M = \frac{\det(R_{\alpha\bar\beta})}{K},$$

where $v_M = i^n K\, dz^1 \wedge d\bar z^1 \wedge \cdots \wedge dz^n \wedge d\bar z^n$ and $\varphi_M = 2i \sum R_{\alpha\bar\beta}\, dz^\alpha \wedge d\bar z^\beta$. For M', we define $r_{M'}$ in the same way.

THEOREM 3.3.   *Let M and M' be n-dimensional complex manifolds with volume elements v_M and $v_{M'}$, respectively. Assume*

(a) *The Ricci tensors h_M and $h_{M'}$ are negative definite;*
(b) *$r_M(p)/r_{M'}(p') \geqq 1$ for $p \in M$ and $p' \in M'$;*
(c) *M' is compact.*

Then every holomorphic mapping $f : M' \to M$ is volume-decreasing in the sense that $f^* v_M / v_{M'} \leqq 1$.

Proof.   We set $u = f^* v_M / v_{M'}$ as before. We have

$$\frac{f^* \varphi_M^n}{\varphi_{M'}^n} = \frac{f^* r_M}{r_{M'}} \frac{f^* v_M}{v_{M'}} = \frac{f^* r_M}{r_{M'}} u \geqq u,$$

where the last inequality is a consequence of (b). Let $p' \in M'$ be a point where u attains its maximum (which may be assumed to be nonzero). To prove that $u \leqq 1$, it suffices to show that $f^* \varphi_M^n / \varphi_{M'}^n \leqq 1$ at p'. But this follows from the inequalities:

$$h_{M'} \leqq f^* h_M < 0 \qquad \text{at} \quad p',$$

where the first inequality is a consequence of Theorem 2.1 and the second inequality follows from (a).   QED.

COROLLARY 3.4.   *Let M and M' be n-dimensional Hermitian Einstein manifolds with metric $ds_M{}^2$ and $ds_{M'}^2$ such that $h_M = -ds_M{}^2$ (i.e., $R_{\alpha\bar\beta} = -g_{\alpha\bar\beta}$) and $h_{M'} = -ds_{M'}^2$. If M' is compact, every holomorphic mapping $f : M' \to M$ is volume-decreasing.*

THEOREM 3.5.   *In Theorem 3.3 or in Corollary 3.4, assume further that M is also compact. Let $V(M)$ and $V(M')$ denote the total volumes of M and M', respectively:*

$$V(M) = \int_M v_M, \qquad V(M') = \int_{M'} v_{M'}.$$

(1) *If $V(M') < V(M)$, then every holomorphic mapping $f : M' \to M$ is degenerate everywhere on M';*

(2) *If $V(M') = V(M)$, then every holomorphic mapping $f : M' \to M$ is either degenerate everywhere on M' or a volume-preserving biholomorphic mapping.*

Proof. The topological degree $\deg f$ of f can be given by

$$\deg f = \frac{1}{V(M)} \int_{M'} f^* v_M.$$

Since $f^* v_M \leq v_{M'}$ by Theorem 3.3, we obtain

$$\deg f \leq V(M')/V(M).$$

Since f is holomorphic, $f^* v_M$ is nonnegative, i.e., $f^* v_M / v_{M'} \geq 0$, and $\deg f \geq 0$. Moreover, $\deg f = 0$ only when $f^* v_M = 0$, i.e., only when f is degenerate everywhere on M'. Since $\deg f$ is always an integer, we may conclude that *$\deg f$ is a nonnegative integer such that* $\deg f \leq V(M')/V(M)$, *and $\deg f = 0$ if and only if f is degenerate everywhere on M'.* If $V(M') < V(M)$, then $\deg f = 0$. This proves (1). If $V(M') = V(M)$, then $f^* v_M \leq v_{M'}$ and

$$0 \leq \deg f = \frac{1}{V(M)} \int_{M'} f^* v_M \leq \frac{V(M')}{V(M)} = 1.$$

Assume that f is not everywhere degenerate. Then

$$f^* v_M = v_{M'}.$$

Being an everywhere nondegenerate mapping of a compact manifold into another compact manifold, $f : M' \to M$ is a covering projection. Since $\deg f = 1$, it follows that f is a biholomorphic mapping.   QED.

COROLLARY 3.6. *Let M be a compact Hermitian Einstein manifold with negative definite Ricci tensor (i.e., $R_{\alpha\bar\beta} = -c g_{\alpha\bar\beta}$ with $c > 0$). Every holomorphic mapping $f : M \to M$ is either degenerate everywhere or biholomorphic and isometric.*

Proof. By Theorem 3.5, if f is nondegenerate somewhere, then f is biholomorphic and volume-preserving. If f is volume-preserving, it preserves the Ricci tensor h_M and also the metric $-(1/c)h_M$.   QED.

The proof of Theorem 3.5 gives also the following:

COROLLARY 3.7. *Let M and M' be compact Hermitian Einstein manifolds such that $h_M = -c \, ds_M{}^2$ and $h_{M'} = -c \, ds^2_{M'}$ with $c > 0$. Then, for every holomorphic mapping $f : M' \to M$, we have*

$$\deg f \leqq V(M')/V(M).$$

Remark. In certain cases, the total volumes $V(M)$ and $V(M')$ may be expressed in terms of topological invariants of M and M'. Assume $h_M = -c \, ds_M{}^2$ and $h_{M'} = -c \, ds^2_{M'}$ as in Corollary 3.7. Since the first Chern classes $c_1(M)$ and $c_1(M')$ are represented by $(1/4\pi)\varphi_M$ and $(1/4\pi)\varphi_{M'}$, we see that $c_1(M)^n$ and $c_1(M')^n$ are represented by $(-1)^n av_M$ and $(-1)^n av_{M'}$, respectively, where a is a positive constant. Hence, the ratio of the volumes may be expressed as the ratio of certain Chern numbers, namely,

$$V(M')/V(M) = c_1{}^n[M']/c_1{}^n[M].$$

Another case that may be of interest is the case where both M and M' are covered by the same universal covering manifold which is a homogeneous Hermitian manifold. Since M and M' are locally homogeneous and isometric, the integrands in the Gauss–Bonnet formulas for M and M' are the same constant. Hence,

$$V(M')/V(M) = \chi(M')/\chi(M),$$

where χ denotes the Euler–Poincaré characteristic.

If we apply the last remark to Riemann surfaces of genus $\geqq 2$, then we may conclude the following:

COROLLARY 3.8.   (1) *If M and M' are compact Riemann surfaces such that*

$$genus \ of \ M > genus \ of \ M' \geqq 2,$$

then every holomorphic mapping $f : M' \to M$ is a constant map;
     (2) *If M and M' are compact Riemann surfaces such that*

$$genus \ of \ M = genus \ of \ M' \geqq 2,$$

then every holomorphic mapping $f : M' \to M$ is either a constant map or a biholomorphic mapping.

This corollary may be derived easily from the nonintegrated form of the second main theorem of Nevanlinna theory. As a matter of fact, Nevanlinna theory implies not only the inequality of Corollary 3.7,

$$\deg f \leq \chi(M')/\chi(M)$$

for compact Riemann surfaces M and M', but also the equality

$$\deg f = \chi(M')/\chi(M) + n_1(M')/\chi(M),$$

where $n_1(M')$ is a certain nonnegative integer called the stationary index of f. For details, see Wu [1, Corollaries 3.2 and 3.3]. A generalization of the equality above to higher dimension is not known.

We conclude this section by a remark related to Corollary 3.6. Let M be a compact complex manifold such that a suitable positive power K^m of its canonical line bundle K admits nontrivial holomorphic cross sections. Then every holomorphic mapping f of M into itself which is nondegenerate at some point is a covering projection (see Peters [1]). This result of Peters is related to Corollary 3.6 as follows. If M is a compact Hermitian manifold with negative definite Ricci tensor [or with negative first Chern class $c_1(M)$], then the line bundle K^m for a suitable positive integer m admits sufficiently many holomorphic sections to define an imbedding of M into a projective space (Kodaira [1]). It should be noted also that if $H^1(M; \mathbf{R}) = 0$ and $c_1(M) = 0$, then K itself admits a holomorphic section which vanishes nowhere, i.e., K is a trivial bundle. Other related results of Peters will be discussed in Chapter VIII.

4. *Holomorphic Mappings $f : D \to M$, Where D Is a Homogeneous Bounded Domain*

Although we have a homogeneous bounded domain D in mind, we shall treat D as an abstract complex manifold satisfying certain conditions.

LEMMA 4.1. *Let D and M be n-dimensional complex manifolds with volume elements v_D and v_M respectively. Let $f : D \to M$ be a holomorphic mapping. Assume*

(a) *The Ricci tensors h_D and h_M are negative definite;*
(b) *$r_M(q)/r_D(p) \geq 1$ for $p \in D$ and $q \in M$;*

(c) *The function $u = f^* v_M / v_D$ tends to zero at the boundary of D in the sense that for every positive number a the set $\{p \in D; u(p) \geq a\}$ is compact.*

Then f is volume-decreasing, i.e., $u \leq 1$.

See § 3 for the definition of r_M and r_D.

Proof. In the proof of Theorem 3.3, the compactness of M' (which corresponds to D here) was needed only to ensure the existence of a point where u attains its maximum. Since (c) guarantees the existence of such a point, Lemma 4.1 follows from the proof of Theorem 3.3.

QED.

LEMMA 4.2. *In Lemma 4.1, f is volume-decreasing if* (c) *is replaced by the following assumptions*:

(c') *there exists a sequence $D_1 \subset D_2 \subset \cdots \subset D$ of open submanifolds such that*

(c'.1) $\cup_k D_k = D$;

(c'.2) *each D_k carries a volume element v_k for which $(D_k, M, f \mid D_k)$ satisfies* (a), (b), *and* (c) *of Lemma 4.1 and moreover*

$$\lim_{k \to \infty} v_k = v_D \qquad \text{(pointwise on } D).$$

Proof. Applying Lemma 4.1 to $f : D_k \to M$, we obtain

$$f^* v_M / v_k \leq 1 \quad \text{on} \quad D_k.$$

Hence,

$$f^* v_M / v_D = \lim_{k \to \infty} f^* v_M / v_k \leq 1.$$

QED.

We shall now apply Lemma 4.2 to the case where D is a bounded domain in $\mathbf{C}^n$. With respect to the natural coordinate system $z^1, \ldots, z^n$ in $\mathbf{C}^n$, the Bergman kernel form b_D can be written as (see Example 2 in § 1)

$$b_D = i^n K(z, \bar{z}) \, dz^1 \wedge d\bar{z}^1 \wedge \cdots \wedge dz^n \wedge d\bar{z}^n,$$

where $K(z, \bar{z})$ is the Bergman kernel function of the domain D.

LEMMA 4.3. *Let D be a bounded domain in $\mathbf{C}^n$ with the Bergman kernel form b_D. Assume that*

(i) *the Bergman kernel function $K(z, \bar{z})$ of D tends to infinity at the boundary of D;*

(ii) *D is starlike in the sense that, for each $0 \leq a < 1$, the set*

$$D_a = \{az \in \mathbf{C}^n; \ z \in D\} \quad (z \text{ being considered as a vector})$$

is contained in D.

If b_a denotes the Bergman kernel form of D_a for $0 < a < 1$, then

$$\lim_{a \to 1} b_a = b_D;$$

and for any complex manifold M of dimension n with volume element v_M and for any holomorphic mapping $f : D \to M$, the function $f^* v_M / b_a$ on D_a tends to zero at the boundary of D_a.

Proof. For each $0 < a < 1$, the mapping $g_a : D \to D_a$ defined by $g_a(z) = az$ is a biholomorphic mapping. Hence, $g_a^*(b_a) = b_D$. If we write

$$b_D = i^n K(z, z) \, dz^1 \wedge d\bar{z}^1 \wedge \cdots \wedge dz^n \wedge d\bar{z}^n$$

and

$$b_a = i^n K_a(z, \bar{z}) \, dz^1 \wedge d\bar{z}^1 \wedge \cdots \wedge dz^n \wedge d\bar{z}^n,$$

then the invariance $g_a^*(b_a) = b_D$ implies

$$K(z, \bar{z}) = a^{2n} K_a(az, a\bar{z}) \quad \text{for} \quad z \in D.$$

It is now clear that $\lim_{a \to 1} b_a = b_D$. If we write

$$f^* v_M = i^n L(z, \bar{z}) \, dz^1 \wedge d\bar{z}^1 \wedge \cdots \wedge dz^n \wedge d\bar{z}^n,$$

then the function $L(z, \bar{z})$ is continuous on D and hence is bounded on the closure $\bar{D}_a \subset D$ for each a. Since $K(z, \bar{z})$ goes to infinity at the boundary of D by assumption, so does $K_a(z, \bar{z})$ at the boundary of D_a. Hence, $L(z, \bar{z})/K_a(z, \bar{z})$ tends to zero at the boundary of D_a. Since $f^* v_M / b_a = L/K_a$, this completes the proof.   QED.

THEOREM 4.4. *Let D be a bounded domain in $\mathbf{C}^n$ with the Bergman kernel form b_D and the volume element v_D defined by the Bergman metric. Assume*

(i) *the Bergman kernel function K of D goes to infinity at the boundary of D;*

(ii) *D is starlike;*

(iii) *the ratio v_D/b_D is a constant function on D.*

Let M be an n-dimensional Hermitian manifold with metric $ds_M^2 = 2 \sum g_{\alpha\bar\beta} \, dw^\alpha \, d\bar{w}^\beta$ and volume element v_M. Assume

(iv) *the Ricci tensor $h_M = 2 \sum R_{\alpha\bar\beta} \, dw^\alpha \, d\bar{w}^\beta$ is negative definite and the associated 2-form $\varphi_M = 2i \sum R_{\alpha\bar\beta} \, dw^\alpha \wedge d\bar{w}^\beta$ satisfies*

$$(-1)^n \varphi_M{}^n / 2^n v_M \geq 1 ,$$

i.e.,

$$(-1)^n \det(R_{\alpha\bar\beta}) / \det(g_{\alpha\bar\beta}) \geq 1 .$$

Then every holomorphic mapping $f : D \to M$ is volume-decreasing, i.e.,

$$f^* v_M / v_D \leq 1 .$$

Proof. Let ds_D^2 be the Bergman metric of D and h_D its Ricci tensor. By (iii) h_D is the Ricci tensor of the volume element b_D as well as that of v_D and hence is equal to $-ds_D^2$. If we define r_D as in § 3, then we have

$$r_D = (-1)^n .$$

Since $(-1)^n r_M \geq 1$ by (iv), we have

$$r_M(q)/r_D(p) \geq 1 \qquad \text{for} \quad p \in D \quad \text{and} \quad q \in M ,$$

which verifies (b) of Lemma 4.1. Since v_D/b_D is a constant function on D, v_a/b_a is also a constant function on D_a. It is therefore clear from Lemma 4.3 that

$$\lim_{a \to 1} v_a = v_D$$

and

$$f^* v_M / v_a \quad \text{tends to zero at the boundary of } D_a ,$$

which verifies (c′) of Lemma 4.2. Since $h_D = -ds_D^2$, h_D is negative definite. On the other hand, h_M is negative definite by assumption. This verifies (a) of Lemma 4.1. Now, Theorem 4.4 follows from Lemma 4.2.   QED.

Remark. If M is a compact complex manifold with ample canonical line bundle, M admits an Hermitian metric satisfying (iv). Let D be a bounded homogeneous domain in $\mathbf{C}^n$. As we remarked in Example 2 of § 1, the ratio v_D/b_D is a constant function on D. By a well-known theorem of Vinberg, Gindikin, and Pyatetzki-Shapiro [1], D is biholomorphic with an affinely homogeneous Siegel domain of second kind. In the next section we shall show that every affinely homogeneous Siegel domain of second kind satisfies (i) and (ii) of Theorem 4.4 as well as (iii). Thus, Theorem 4.4 may be applied to every bounded homogeneous domain D in $\mathbf{C}^n$ and, in particular, to every Hermitian symmetric space D of noncompact type.

5. *Affinely Homogeneous Siegel Domains of Second Kind*

Following Pyateztki-Shapiro [1] we define the Siegel domains of second kind. Let V be a convex cone in $\mathbf{R}^n$, containing no entire straight lines. A mapping $F : \mathbf{C}^m \times \mathbf{C}^m \to \mathbf{C}^n$ is said to be *V-Hermitian* if

(1) $F(u, v) = \overline{F(v, u)}$ for $u, v \in \mathbf{C}^m$,

(2) $F(au_1 + bu_2, v) = a\, F(u_1, v) + b\, F(u_2, v)$ for $u_1, u_2, v \in \mathbf{C}^m$, $a, b \in \mathbf{C}$,

(3) $F(u, u) \in \overline{V}$ (the closure of V)   for $u \in \mathbf{C}^m$,

(4) $F(u, u) = 0$ only when $u = 0$.

The subset S of $\mathbf{C}^{n+m}$ defined by

$$S = \{(z, u) \in \mathbf{C}^n \times \mathbf{C}^m; \ \operatorname{Im} z - F(u, u) \in V\}$$

is called the *Siegel domain of second kind defined by V and F*. Before we prove that S is equivalent to a bounded domain in $\mathbf{C}^{n+m}$, we consider the Siegel domain of second kind in $\mathbf{C}^{1+m}$ defined by

$$\operatorname{Im} z - |u^1|^2 - \cdots - |u^m|^2 > 0,$$

where $z, u^1, \ldots, u^m$ are the coordinate functions in $\mathbf{C}^{1+m}$. We prove first this domain is biholomorphic with the unit ball

$$|z^0|^2 + \cdots + |z^m|^2 < 1$$

in $\mathbf{C}^{1+m}$. We set

$$z^0 = \frac{z - i}{z + i}, \qquad z^1 = \frac{2u^1}{z + i}, \qquad \ldots, \qquad z^m = \frac{2u^m}{z + i}.$$

Then we have

$$1 - \sum_{k=0}^{m} | z^k |^2 = \frac{4}{| z + i |^2} (\mathrm{Im}\, z - | u^1 |^2 - \cdots - | u^m |^2),$$

which proves our assertion.

We shall now prove, following Pyateztki-Shapiro, that the Siegel domain S of second kind defined by V and F is equivalent to a domain contained in a product of balls. By a linear change of coordinate system, we may always assume that V is contained in the cone $y^1 > 0, \ldots, y^n > 0$. If V is the cone $y^1 > 0, \ldots, y^n > 0$, then the components $F^1(u, u), \ldots,$ $F^n(u, u)$ are positive semidefinite Hermitian form in $u^1, \ldots, u^m$. We represent each $F^k(u, u)$ as a sum of squares of linear forms:

$$F^k(u, u) = | L_1^k |^2 + \cdots + | L_{s_k}^k |^2.$$

We define new Hermitian forms $\tilde{F}^1, \ldots, \tilde{F}^n$ as follows. We set

$$\tilde{F}^1(u, u) = F^1(u, u)$$
$$\tilde{F}^2(u, u) = {\sum_s}' | L_s^2 |^2,$$

where the prime indicates that the summation is restricted to those L_s^2 that are not linear combinations of $L_1^1, \ldots, L_{s_1}^1$. Then we set

$$\tilde{F}^3(u, u) = {\sum_s}' | L_s^3 |^2,$$

where the prime indicates that the summation is restricted to those L_s^3 that are not linear combinations of $L_1^1, \ldots, L_{s_1}^1, L_1^2, \ldots, L_{s_2}^2$. Similarly, we define $\tilde{F}^4, \ldots, \tilde{F}^n$. Let $\tilde{S}$ be the domain defined by

$$\mathrm{Im}\, z^k - \tilde{F}^k(u, u) > 0 \qquad k = 1, \ldots, n.$$

Since $F^k(u, u) \geq \tilde{F}^k(u, u)$ for all k, it follows that $\tilde{S}$ contains S. We shall show that $\tilde{S}$ is equivalent to a product of balls. By (4) in the definition of F, the system of linear equations

$$L_s^k = 0 \qquad k = 1, \ldots, n, \quad s = 1, \ldots, s_k$$

has only one solution, $u = 0$. It follows that the number of the linear forms L_s^k which really appeared in the construction of $\tilde{F}^1, \ldots, \tilde{F}^k$ is equal to m. By construction, these forms are linearly independent. If we take them as new variables $v^1, \ldots, v^m$, then the domain $\tilde{S}$ is defined by

$$\operatorname{Im} z^1 - |v^1|^2 - \cdots - |v^{m_1}|^2 > 0,$$
$$\operatorname{Im} z^2 - |v^{m_1+1}|^2 - \cdots - |v^{m_2}|^2 > 0,$$
$$\cdots\cdots\cdots\cdots\cdots\cdots$$
$$\operatorname{Im} z^n - |v^{m_{n-1}+1}|^2 - \cdots - |v^m|^2 > 0.$$

It follows that $\tilde{S}$ is equivalent to a product of balls in $\mathbf{C}^{1+m_1}$, $\mathbf{C}^{1+m_2-m_1}$, $\ldots, \mathbf{C}^{1+m-m_{n-1}}$.

The following proposition is due to Hahn and Mitchell [2].

PROPOSITION 5.1. *Let S be the Siegel domain of second kind in $\mathbf{C}^{n+m}$ defined by V and F. Let t be a fixed point of V and set*

$$S_k = \left\{ (z, u) \in \mathbf{C}^n \times \mathbf{C}^m; \operatorname{Im} z - \frac{t}{k} - F(u, u) \in V \right\} \text{ for } k=1, 2, 3, \ldots.$$

Then $S_1 \subset S_2 \subset S_3 \subset \cdots \subset S$, $\cup_k S_k = S$ and $\bar{S}_k \subset S_{k+1}$. For each k, the translation

$$(z, u) \in S \rightarrow \left(z + \frac{t}{k}, u \right) \in S_k$$

gives a biholomorphic mapping from S onto S_k.

The proof is straightforward and hence is omitted.

For the Siegel domain S of second kind defined by V and F, an *automorphism* of V is a linear transformation A of V such that, for a suitable complex linear transformation B of $\mathbf{C}^m$, the following holds:

$$A F(u, u) = F(Bu, Bu) \qquad \text{for } u \in \mathbf{C}^m.$$

Then, for any $x_0 \in \mathbf{R}^n$ and $u_0 \in \mathbf{C}^m$, the affine transformation

$$z \rightarrow Az + x_0 + 2i F(Bu, u_0) + i F(u_0, u_0)$$
$$u \rightarrow Bu + u_0$$

is an automorphism of the Siegel domain S. If V is homogeneous, i.e., the group of automorphisms A of V is transitive, then S is homogeneous under the group of holomorphic transformations (which are affine in $\mathbf{C}^{n+m}$). In fact, if $y_0 \in V$ so that $(iy_0, 0) \in S$ and if (z_1, u_1) is an arbitrary point of S so that $y_1 = \operatorname{Im} z_1 - F(u_1, u_1) \in V$, then the transformation

$$z \to Az + \operatorname{Re} z_1 + 2i\, F(Bu, u_1) + i\, F(u_1, u_1)$$
$$u \to Bu + u_1$$

maps $(iy_0, 0)$ into (z_1, u_1), where A is an automorphism of V sending y_0 into y_1 and B is a linear map of $\mathbf{C}^m$ satisfying $A\, F(u, u) = F(Bu, Bu)$. The Siegel domain S of second kind defined by V and F is said to be *affinely homogeneous* if V is homogeneous. A theorem of Vinberg, Gindikin, and Pyatetzki-Shapiro [1] says that every homogeneous bounded domain in $\mathbf{C}^n$ is biholomorphic to an affinely homogeneous Siegel domain of second kind.

PROPOSITION 5.2. *Let D be a homogeneous bounded domain in $\mathbf{C}^n$ and $K(z, \bar{z})$ its Bergman kernel function. Then $K(z, \bar{z})$ goes to infinity at the boundary of D.*

Proof. Assume that the proposition is false. Then there exists a sequence of points $z_1, z_2, \ldots$ in D which converges to a point on the boundary of D such that

$$K(z_k, \bar{z}_k) \leq a \qquad \text{for} \quad k = 1, 2, 3, \ldots,$$

where a is a positive constant. Fix a point z_0 of D. For each k, let f_k be an automorphism of D such that

$$f_k(z_0) = z_k.$$

Let J_k denote the Jacobian of f_k at z_0 (with respect to the Euclidean coordinate system in $\mathbf{C}^n$). Since the Bergman kernel form of D is invariant by f_k, we obtain

$$K(z_0, \bar{z}_0) \;\; = K(z_k, \bar{z}_k)\,|J_k|^2 \qquad \text{for} \quad k = 1, 2, 3, \ldots.$$

It follows that

$$|J_k| \geqq b > 0 \qquad \text{for} \quad k = 1, 2, 3, \ldots.$$

Consider $\{f_k\}$ as a family of holomorphic mappings from D into $\mathbf{C}^n$ which are uniformly bounded. Taking a subsequence if necessary, we may assume that $\{f_k\}$ converges to a holomorphic mapping $f : D \to \mathbf{C}^n$

uniformly on all compact subsets of D. It is well known that uniform convergence of $\{f_k\}$ entails uniform convergence of all corresponding partial derivatives of all orders of $\{f_k\}$. In particular, the Jacobian $Jf(z_0)$ of f at z_0 satisfies

$$|Jf(z_0)| = \lim |J_k| \geqq b > 0.$$

Hence, f gives a homeomorphism of a neighborhood $U \subset D$ of z_0 onto the neighborhood $f(U)$ of $f(z_0)$ in $\mathbf{C}^n$. Since $f(z_0)$ is a boundary point of D, $f(U)$ is not contained in $\bar{D}$. On the other hand, being the limit of $\{f_k\}$, f maps D into $\bar{D}$. This is a contradiction.   QED.

Proposition 5.2 shows that an affinely homogeneous Siegel domain of the second kind satisfies (i) of Theorem 4.4. Instead of (ii) in Theorem 4.4, we use Proposition 5.1. Now, from Theorem 4.4, we obtain

THEOREM 5.3. *Let D be an n-dimensional complex manifold which is biholomorphic with an affinely homogeneous Siegel domain of second kind and let $ds_D{}^2$ be its Bergman metric. Let M be an n-dimensional Einstein–Kaehler manifold with metric $ds_M{}^2$, such that its Ricci tensor h_M is equal to $-ds_M{}^2$. Then every holomorphic mapping $f : D \to M$ is volume-decreasing.*

If we assume the result of Vinberg, Gindkin, and Pyatetzki-Shapiro, then in Theorem 5.3 D can be any homogeneous bounded domain in $\mathbf{C}^n$.

Dinghas [1] obtained Theorem 5.3 when D is a unit ball in $\mathbf{C}^n$. A result similar to Lemma 4.2 was proved by Chern [1] under stronger curvature assumptions. Hahn and Mitchell [1] and Kobayashi [2] proved Theorem 5.3 independently when D is a symmetric bounded domain. The generalization to the case where D is a homogeneous bounded domain is due to Hahn and Mitchell [2]. But in both papers Hahn and Mitchell assumed unnecessarily that f is biholomorphic.

6. *Symmetric Bounded Domains*

We shall first summarize known results on the Bergman kernel functions of the so-called Cartan classical domains. For details we refer the reader to Hua [1], Tashiro [1], and Hahn and Mitchell [1].

According to E. Cartan [1], there exist only six types of irreducible bounded symmetric domains—four classical types and two exceptional types. The four classical domains R_I, R_{II}, R_{III}, and R_{IV} are defined as follows:

$R_{\rm I}$ = $\{m \times n$ matrices Z satisfying $I_m - ZZ^* > 0\}$,

$R_{\rm II}$ = $\{$symmetric matrices Z of order n satisfying $I_n - ZZ^* > 0\}$,

$R_{\rm III}$ = $\{$skew-symmetric matrices Z of order n satisfying $I_n - ZZ^* > 0\}$,

$R_{\rm IV}$ = $\{z = (z_1, \ldots, z_n) \in \mathbf{C}^n;\ |zz'|^2 + 1 - 2\bar{z}z' > 0,\ |zz'| < 1\}$,

where I_m denotes the identity matrix of order m, Z^* is the complex conjugate of the transposed Z' of Z, and z' is the transposed of the vector z.

For the domain R_j, $(j = {\rm I, II, III, IV})$, we denote by K_j, b_j, $ds_j{}^2$ and v_j the Bergman kernel function, the Bergman kernel form, the Bergman metric, and the volume element defined by $ds_j{}^2$, respectively. The ratio v_j/b_j will be denoted by c_j. We denote by $V(R_j)$ the total volume of R_j with respect to the Euclidean measure of ambient complex Euclidean space. Then

$$V(R_{\rm I}) = \frac{1!2! \ldots (m-1)!1!2! \ldots (n-1)!}{1!2! \cdots (m+n-1)!} \pi^{mn},$$

$$V(R_{\rm II}) = \frac{2!4! \cdots (2n-2)!}{n!(n+1)! \cdots (2n-1)!} \pi^{n(n+1)/2},$$

$$V(R_{\rm III}) = \frac{2!4! \cdots (2n-4)!}{(n-1)!n! \cdots (2n-3)!} \pi^{n(n-1)/2},$$

$$V(R_{\rm IV}) = \frac{1}{2^{n-1}n!} \pi^n.$$

$$K_{\rm I}(Z, \bar{Z}) = \frac{1}{V(R_{\rm I})} \{\det(I_m - ZZ^*)\}^{-(m+n)},$$

$$K_{\rm II}(Z, \bar{Z}) = \frac{1}{V(R_{\rm II})} \{\det(I_n - ZZ^*)\}^{-(n+1)},$$

$$K_{\rm III}(Z, Z) = \frac{1}{V(R_{\rm III})} \{\det(I_n - ZZ^*)\}^{-(n-1)},$$

$$K_{\rm IV}(z, \bar{z}) = \frac{1}{V(R_{\rm IV})} (|zz'|^2 + 1 - 2\bar{z}z')^{-n}.$$

$$ds_{\rm I}{}^2 = 2(m+n)\, {\rm Trace}[(I_m - ZZ^*)^{-1}\, dZ(I_n - Z^*Z)^{-1}\, dZ^*],$$

$$ds_{\rm II}^2 = 2(n+1)\, {\rm Trace}[(I_n - ZZ^*)^{-1}\, dZ(I_n - Z^*Z)^{-1}\, dZ^*],$$

$$ds_{\rm III}^2 = 2(n-1)\, {\rm Trace}[(I_n - ZZ^*)^{-1}\, dZ(I_n - Z^*Z)^{-1}\, dZ^*],$$

$$ds_{\rm IV}^2 = 4nA\, dz[A(I_n - z'\bar{z}) + (I_n - z'\bar{z})z^*z(I_n - z'\bar{z})]\, dz^*,$$

where

$$A = |zz'|^2 + 1 - 2\bar{z}z',$$
$$c_{\mathrm{I}} = (m + n)^{mn}\, V(R_{\mathrm{I}}),$$
$$c_{\mathrm{II}} = 2^{n(n-1)/2}(n + 1)^{n(n+1)/2}\, V(R_{\mathrm{II}}),$$
$$c_{\mathrm{III}} = [2(n - 1)]^{n(n-1)/2}\, V(R_{\mathrm{III}}),$$
$$c_{\mathrm{IV}} = (2n)^n\, V(R_{\mathrm{IV}}).$$

If we denote by G_j the automorphism group of the domain R_j, then a transformation $Z \in R_j \to W \in R_j$ belonging to G_j is given as follows.

For R_{I} we have

$$W = (AZ + B)(CZ + D)^{-1},$$

where A, B, C, D are matrices of dimensions $m \times m$, $m \times n$, $n \times m$, and $n \times n$, respectively, satisfying the relations

$$AA^* - BB^* = I_m, \qquad AC^* = BD^*, \qquad CC^* - DD^* = -I_n.$$

For R_{II} we have

$$W = (AZ + B)(\bar{B}Z + \bar{A})^{-1},$$

where

$$A'B = B'A, \qquad AA^* - BB^* = I_n.$$

For R_{III} we have

$$W = (AZ + B)(-\bar{B}Z + \bar{A})^{-1},$$

where

$$A'B = -B'A, \qquad A^*A - B^*B = I_n.$$

For R_{IV} we have

$$w = \left\{ \left[\left(\frac{1}{2}(zz' + 1), \frac{i}{2}(zz' - 1) \right) A' + zB' \right] \binom{1}{i} \right\}^{-1}$$
$$\times \left\{ \left(\frac{1}{2}(zz' + 1), \frac{i}{2}(zz' - 1) \right) C' + zD' \right\},$$

where A, B, C, and D are real matrices of dimensions 2×2, $2 \times n$,

$n \times 2$, and $n \times n$, respectively, satisfying the relations

$$\begin{pmatrix} A & B \\ C & D \end{pmatrix} \begin{pmatrix} I_2 & 0 \\ 0 & -I_n \end{pmatrix} \begin{pmatrix} A & B \\ C & D \end{pmatrix}' = \begin{pmatrix} I_2 & 0 \\ 0 & -I_n \end{pmatrix}, \qquad \det\begin{pmatrix} A & B \\ C & D \end{pmatrix} = 1$$

Let $Z \in R_j \to W \in R_j$ be a holomorphic mapping from the domain R_j into itself and let $J(Z)$ be its Jacobian. By Theorem 4.4 we have the following inequalities:

For $j =$ I,

$$\begin{aligned} |J(Z)|^2 &\leq \{\det(I_m - WW^*)/\det(I_m - ZZ^*)\}^{m+n} \\ &\leq \{\det(I_m - ZZ^*)\}^{-(m+n)}. \end{aligned}$$

For $j =$ II,

$$\begin{aligned} |J(Z)|^2 &\leq \{\det(I_n - WW^*)/\det(I_n - ZZ^*)\}^{n+1} \\ &\leq \{\det(I_n - ZZ^*)\}^{-(n+1)}. \end{aligned}$$

For $j =$ III,

$$\begin{aligned} |J(Z)|^2 &\leq \{\det(I_n - WW^*)/\det(I_n - ZZ^*)\}^{n-1} \\ &\leq \{\det(I_n - ZZ^*)\}^{-(n-1)}. \end{aligned}$$

For $j =$ IV,

$$\begin{aligned} |J(z)|^2 &\leq \{(|ww'|^2 + 1 - 2\bar{w}w')/(|zz'|^2 + 1 - 2\bar{z}z')\}^n \\ &\leq (|zz'|^2 + 1 - 2\bar{z}z')^{-n}. \end{aligned}$$

III

Distance and the Schwarz Lemma

1. Hermitian Vector Bundles and Curvatures

We shall summarize basic local formulas of Hermitian differential geometry. For a fuller treatment, see Kobayashi and Nomizu [1; Vol. II].

Let E be a holomorphic vector bundle over a complex manifold M with fiber $\mathbf{C}^r$. We denote by E_p the fiber of E at $p \in M$. A *Hermitian fiber-metric* g assigns to each point $p \in M$ a Hermitian inner product g_p in E_p. A holomorphic vector bundle E with a Hermitian fiber-metric g is called a *Hermitian vector bundle*.

For a local coordinate system $z^1, \ldots, z^n$ of M, we set

$$Z_i = \partial/\partial z^i, \qquad Z_{\bar{i}} = \bar{Z}_i = \partial/\partial \bar{z}^i \qquad \text{for} \quad i = 1, \ldots, n.$$

Let $e_1, \ldots, e_r$ be holomorphic local cross sections of E which are everywhere linearly independent. With respect to this holomorphic local frame field, the components $g_{\alpha\bar{\beta}}$ of g are given by

$$g_{\alpha\bar{\beta}} = g(e_\alpha, \bar{e}_\beta) \qquad \text{for} \quad \alpha, \beta = 1, \ldots, r.$$

The components of the Hermitian connection are given by

$$\Gamma_{i\beta}^\alpha = \sum_\gamma g^{\alpha\bar{\gamma}} \frac{\partial g_{\beta\bar{\gamma}}}{\partial z^i}, \qquad \Gamma_{\bar{i}\bar{\beta}}^{\bar{\alpha}} = \bar{\Gamma}_{i\beta}^\alpha.$$

All the other components are set equal to zero. The covariant derivatives $\nabla_{Z_i} e_\beta$ are defined by

$$\nabla_{Z_i} e_\beta = \sum_\alpha \Gamma_{i\beta}^\alpha e_\alpha.$$

The components $K^\alpha_{\beta ij}$ of the curvature R are defined by

$$\sum_\alpha K^\alpha_{\beta ij} e_\alpha = R(Z_i, Z_j) e_\beta = ([\nabla_{Z_i}, \nabla_{Z_j}] - \nabla_{[Z_i, Z_j]}) e_\beta.$$

Then

$$\sum_\alpha K^\alpha_{\beta ij} = -\frac{\partial \Gamma^\alpha_{i\beta}}{\partial \bar{z}^j} = -\sum_\gamma g^{\alpha \bar\gamma} \frac{\partial^2 g_{\beta \bar\gamma}}{\partial z^i \, \partial \bar{z}^j} + \sum_{\gamma, \delta, \epsilon} g^{\alpha \bar\delta} g^{\epsilon \bar\gamma} \frac{\partial g_{\beta \bar\gamma}}{\partial z^i} \frac{\partial g_{\epsilon \bar\delta}}{\partial \bar{z}^j}.$$

If we set

$$K_{\alpha \bar\beta i \bar\delta} = g[R(Z_i, Z_\delta) \bar{e}_\beta, e_\alpha],$$

then

$$K_{\alpha \bar\beta ij} = \frac{\partial^2 g_{\alpha \bar\beta}}{\partial z^i \, \partial \bar{z}^j} - \sum_{\gamma, \epsilon} g^{\epsilon \bar\gamma} \frac{\partial g_{\alpha \bar\gamma}}{\partial z^i} \frac{\partial g_{\epsilon \bar\beta}}{\partial \bar{z}^j}.$$

The components of the Ricci curvature are given by

$$K_{ij} = \sum_\alpha K^\alpha_{\alpha ij} = -\sum_{\alpha, \beta} g^{\alpha \bar\beta} K_{\alpha \bar\beta ij}.$$

If we set

$$G = \det(g_{\alpha \bar\beta}),$$

then

$$\begin{aligned}
K_{ij} &= -\frac{1}{G} \frac{\partial^2 G}{\partial z^i \, \partial \bar{z}^j} + \frac{1}{G^2} \frac{\partial G}{\partial z^i} \frac{\partial G}{\partial \bar{z}^j} \\
&= -\frac{\partial^2 \log G}{\partial z^i \, \partial \bar{z}^j}.
\end{aligned}$$

Given an element $s = \sum s^\alpha e_\alpha$ of E_p, we consider the Hermitian form

$$K_s = -\sum_{\alpha, \beta, i, j} K_{\alpha \bar\beta ij} s^\alpha \bar{s}^\beta \, dz^i \, d\bar{z}^j$$

at $p \in M$. If K_s is positive (resp. negative) definite for every nonzero s, the Hermitian vector bundle E is said to *have positive* (resp. *negative*) *curvature*. If the Hermitian form

$$2 \sum K_{ij} \, dz^i \, d\bar{z}^j$$

is positive (resp. negative) definite, then E is said to *have positive* (resp. *negative*) *Ricci curvature* (or *tensor*).

Let E' be a holomorphic subbundle of E with fiber $\mathbf{C}^q$. With respect to the Hermitian fiber-metric induced by g, E' is also a Hermitian vector bundle. The theory of Hermitian vector subbundles is essentially the same as that of Riemannian submanifolds. We shall give the equation of Gauss which relates the curvature of E' with that of E. In choosing local holomorphic cross sections $e_1, \ldots, e_r$ of E we may assume that $e_1, \ldots, e_q$ are cross sections of E'. If we fix a point p of M, then we may further assume that $e_1, \ldots, e_r$ are orthonormal at p, i.e., $(g_{\alpha\bar\beta})_p = (g(e_\alpha, \bar{e}_\beta))_p = \delta_{\alpha\beta}$. If we denote the curvature of E' by $K'_{\alpha\bar\beta ij}$, then

$$-(K'_{\alpha\bar\beta ij})_p = -(K_{\alpha\bar\beta ij})_p - \sum_{\varrho=q+1}^{r} \left(\frac{\partial g_{\alpha\bar\varrho}}{\partial z^i} \frac{\partial g_{\varrho\bar\beta}}{\partial \bar z^j} \right)_p$$

$$\text{for} \quad \alpha, \beta = 1, \ldots, q; \quad i, j = 1, \ldots, n.$$

If $s = \sum_{\alpha=1}^{q} s^\alpha e_\alpha$ is an element of $E_p' \subset E_p$, then

$$K_s' = K_s - \sum_{i,j} \sum_{\alpha,\beta,\varrho,\sigma} g^{\sigma\bar\varrho} \frac{\partial g_{\alpha\bar\varrho}}{\partial z^i} \frac{\partial g_{\sigma\bar\beta}}{\partial \bar z^j} s^\alpha \bar s^\beta \, dz^i \, d\bar z^j.$$

This implies the inequality

$$K_s' \leq K_s,$$

i.e., $K_s - K_s'$ is a positive semidefinite Hermitian form.

If M is a Hermitian manifold and E is the tangent bundle of M, then the connection considered above is nothing but the classical Hermitian connection. If s, $t \in E_p$ are unit vectors, then $K_s(t, \bar t) = -\sum K_{i\bar\delta kl} s^i \bar s^j t^k \bar t^l$ is what is called the *holomorphic bisectional curvature* determined by s and t in Goldberg and Kobayashi [1]. In particular, $K_s(s, \bar s)$ is called the *holomorphic sectional curvature* determined by s. If M' is a complex submanifold of M, then the tangent bundle $E' = T(M')$ of M' is a Hermitian vector subbundle of $E \mid M'$. Then the formula $K_s' \leq K_s$ above implies a similar inequality for holomorphic bisectional curvature (and hence for holomorphic sectional curvature also).

In summary we state

THEOREM 1.1.   (1) *If E is a Hermitian vector bundle over a complex manifold M and E' is a Hermitian vector subbundle of E, then*

$$K_s' \leq K_s \quad \text{for} \quad s \in E';$$

(2) *If M' is a complex submanifold of a Hermitian manifold M, then the holomorphic bisectional (or sectional) curvature of M' does not exceed that of M.*

2. The Case Where the Domain Is a Disk

We prove a generalization of Theorem 2.1 of Chapter I. Let D_a be the open disk $\{z \in \mathbf{C}; \ |z| < a\}$ of radius of a and let $ds_a{}^2$ be the metric defined by

$$ds_a{}^2 = \frac{4a^2 \, dz \, d\bar{z}}{A(a - |z|^2)^2}.$$

THEOREM 2.1. *Let D_a be the open disk of radius a with the metric $ds_a{}^2$ and let M be an n-dimensional Hermitian manifold whose holomorphic sectional curvature is bounded above by a negative constant $-B$. Then every holomorphic mapping $f : D_a \to M$ satisfies*

$$f^*(ds_M{}^2) \leqq \frac{A}{B} \, ds_a{}^2.$$

Proof. We define a function u on D_a by setting

$$f^*(ds_M{}^2) = u \, ds_a{}^2$$

and want to prove $u \leqq A/B$ everywhere on D_a. As we have shown in the proof of Theorem 2.1 of Chapter I, we may assume that u attains its maximum at a point, say z_0, of D_a. We want to prove that $u \leqq A/B$ at z_0. If $u(z_0) = 0$, then $u \equiv 0$ and there is nothing to prove. Assume that $u(z_0) > 0$. Then the mapping $f : D_a \to M$ is nondegenerate in a neighborhood of z_0 so that f gives a holomorphic imbedding of a neighborhood U of z_0 into M. By Theorem 1.1, the holomorphic sectional curvature of the one-dimensional complex submanifold $f(U)$ of M is bounded above by $-B$. Since $\dim f(U) = 1$, the holomorphic sectional curvature of $f(U)$ is nothing but the Gaussian curvature. The rest of the proof is exactly the same as that of Theorem 2.1 of Chapter I.   QED.

Theorem 2.1 is essentially equivalent to "Aussage 3" in Grauert and Reckziegel [1], in which they assume that the curvature of every one-dimensional complex submanifold of M is bounded above by $-B$.

3. The Case Where the Domain Is a Polydisk

Let $D = D_a{}^l = D_a \times \cdots \times D_a$ be the direct product of l copies of disk D_a of radius a. Let $ds_D{}^2$ be the product metric $ds_a{}^2 + \cdots + ds_a{}^2$ in D. Since $ds_a{}^2$ has constant curvature $-A$, the holomorphic sectional curvature of $ds_D{}^2$ varies between $-A$ and $-A/l$.

THEOREM 3.1.   *Let $D = D_a{}^l$ be a polydisk of dimension l with metric $ds_D{}^2 = ds_a{}^2 + \cdots + ds_a{}^2$ and let M be an n-dimensional Hermitian manifold whose holomorphic sectional curvature is bounded above by a negative constant $-B$. Then every holomorphic mapping $f : D \to M$ satisfies*

$$f^*(ds_M{}^2) \leqq \frac{A}{B}\, ds_D{}^2.$$

Proof.   Let $(r_1, \ldots, r_l)$ be an l-tuple of complex numbers such that $\sum_{i=1}^{l} |r_i|^2 = 1$. Let $j : D_a \to D$ be the imbedding defined by

$$j(z) = (r_1 z, \ldots, r_l z).$$

From the explicit expression of $ds_a{}^2$ given in § 2, we see that j is isometric at the origin of D_a. Let X be a tangent vector of D at the origin. For a suitable $(r_1, \ldots, r_l)$ we can find a tangent vector Y of D_a at the origin such that $j_*(Y) = X$. Then, for any holomorphic mapping $f : D \to M$, we have

$$\| f_* X \|^2 = \| f_* j_* Y \|^2 \leq \frac{A}{B} \| Y \|^2 = \frac{A}{B} \| X \|^2,$$

where the inequality in the middle follows from Theorem 2.1 (applied to $f \circ j : D_a \to M$) and the last equality follows from the fact that j is isometric at the origin of D_a. Since D is homogeneous, the inequality $\| f_* X \|^2 \leqq (A/B) \| X \|^2$ holds for all tangent vectors X of D.   QED.

4. *The Case Where D Is a Symmetric Bounded Domain*

Let D be a symmetric bounded domain of rank l. With respect to a canonical Hermitian metric, its holomorphic sectional curvature lies between $-A$ and $-A/l$ for a suitable positive constant A. For every tangent vector X of D, there is a totally geodesic complex submanifold $D_a{}^l = D_a \times \cdots \times D_a$ (a polydisk of dimension l) such that X is tangent to $D_a{}^l$. (If we write $D = G/K$ and $\mathfrak{g} = \mathfrak{k} + \mathfrak{p}$ in the usual manner and denote by $\mathfrak{a}$ a maximal Abelian subalgebra contained in $\mathfrak{p}$, then $\dim \mathfrak{a} = \operatorname{rank} D = l$. Without loss of generality, we may assume that X is an element of $\mathfrak{a}$ under the usual identification of $\mathfrak{p}$ with the tangent space of D at the origin. If $J : \mathfrak{p} \to \mathfrak{p}$ is the complex structure tensor, then the manifold generated by $\mathfrak{a} + J\mathfrak{a}$ is the desired totally geodesic submanifold $D_a{}^l$. To see this we have to use the fact that, for suitable root vectors X_{α_i} and $X_{-\alpha_i}$ with $i = 1, \ldots, l$, $\mathfrak{a}$ is spanned by $X_{\alpha_i} + X_{-\alpha_i}$, $i = 1, \ldots, l$. But this is rather technical and will not be discussed any further. We shall show later explicitly what $D_a{}^l$ is for each of the classical Cartan domains.)

From Theorem 3.1 we obtain

THEOREM 4.1. *Let D be a bounded symmetric domain with a canonical invariant metric $ds_D{}^2$ whose holomorphic sectional curvature is bounded below by a negative constant $-A$. Let M be a Hermitian manifold with metric $ds_M{}^2$ whose holomorphic sectional curvature is bounded above by a negative constant $-B$. Then every holomorphic mapping $f : D \to M$ satisfies*

$$f^*(ds_M{}^2) \leqq \frac{A}{B}\, ds_D{}^2.$$

COROLLARY 4.2. *Let D be a symmetric bounded domain with a canonical invariant metric $ds_D{}^2$ whose holomorphic sectional curvature is bounded below by a negative constant $-A$. Let M be a symmetric bounded domain of rank l so that its holomorphic sectional curvature lies between $-lB$ and $-B$. Then every holomorphic mapping $f : D \to M$ satisfies*

$$f^*(ds_M{}^2) \leqq \frac{A}{B}\, ds_D{}^2.$$

COROLLARY 4.3. *Let D be a symmetric bounded domain of rank l so that its holomorphic sectional curvature lies between $-lB$ and $-B$. Then every holomorphic mapping $f : D \to D$ satisfies*

$$f^*\, ds_D{}^2 \leqq l\, ds_D{}^2.$$

Both Corollaries 4.2 and 4.3 have been obtained by Koranyi [1].

We shall now exhibit for each of the Cartan domains a totally geodesic polydisk of dimension l ($=$ the rank of the domain).

$R_{\mathrm{I}} = \{m \times n$ matrices Z satisfying $I - ZZ^* > 0\}$,
         rank $= \min(m, n)$.
   $D^l = \{Z = (z_{ij}); z_{ij} = 0$ for $i \neq j\}$, $l = \min(m, n)$.
$R_{\mathrm{II}} = \{$symmetric matrices Z of order n satisfying $I - ZZ^* > 0\}$,
         rank $= n$.
   $D^n = \{Z = (z_{ij}); z_{ij} = 0$ for $i \neq j\}$.

$R_{III} = \{$skew-symmetric matrices Z of order n satisfying $I - ZZ^* > 0\}$,
      rank $= [\frac{1}{2}n]$.

$\quad D^l = \{Z = (z_{ij}); z_{ij} = 0$ except $z_{12} = -z_{21}, z_{34} = -z_{43}, \ldots \}$,
      $l = [\frac{1}{2}n]$.

$R_{IV} = \{z = (z_1, \ldots, z_n) \in \mathbf{C}^n; |zz'|^2 + 1 - 2z\bar{z}' > 0, |zz'| < 1\}$,
      rank $= 2$.

$\quad D^2 = \{(z_1, z_2, 0, \ldots, 0) \in R_{IV}\}$,

where the right-hand side is biholomorphically mapped onto D^2 by

$$(z_1, z_2, 0, \ldots, 0) \to (z_1 + iz_2, z_1 - iz_2).$$

5. Remarks on Holomorphic Mappings of Vector Bundles

Let L (resp. L') be a Hermitian line bundle over a complex manifold M (resp. M') with Hermitian fiber metric g (resp. g'). Let e (resp. e') be a nonvanishing holomorphic local cross section of L (resp. L'). Let $z^1, \ldots, z^n$ (resp. $w^1, \ldots, w^m$) be a local coordinate system in M (resp. M'). If we set

$$G = g(e, \bar{e}) \qquad \text{(resp. } (G' = g'(e', \bar{e}'))$$

$$K_{ij} = -\partial^2 \log G / \partial z^i \, \partial \bar{z}^j \qquad \text{(resp. } K'_{\alpha\beta} = -\partial^2 \log G' / \partial w^\alpha \, \partial \bar{w}^\beta),$$

then

$$h_M = 2 \sum K_{ij} \, dz^i \, d\bar{z}^j \qquad \text{(resp. } h_{M'} = 2 \sum K'_{\alpha\beta} \, dw^\alpha \, d\bar{w}^\beta)$$

is the Ricci tensor of the Hermitian line bundle L (resp. L').

Let $f : L \to L'$ be a fiber-preserving holomorphic mapping which is complex linear on each fiber. We may allow f to vanish at some points. We denote by $u = f^*g/g'$ the function on M' defined by

$$u = f^*g/g' = g(fe', fe')/g'(e', e').$$

The function u does not depend on the choice of e'. If we denote by f the holomorphic map $M' \to M$ induced by $f : L' \to L$, then the complex Hessian of $\log u$ is given by

$$2 \sum \frac{\partial^2 \log u}{\partial w^\alpha \, \partial \bar{w}^\beta} \, dw^\alpha \, d\bar{w}^\beta = h_{M'} - f^*h_M.$$

Of course, this formula makes sense only at those points of M' where $u \neq 0$, i.e., where f is nonzero. To prove the formula, we observe first that

$$u = J \cdot \bar{J} \cdot f^* G / G',$$

where J is a holomorphic function defined by

$$f e'_{p'} = J(p') e_{f(p')} \qquad \text{for} \quad p' \in M'.$$

Since the complex Hessians of $\log J$ and $\log \bar{J}$ vanish, the proof of the formula is essentially the same as that of Theorem 2.1 of Chapter II.

It is now clear that Proposition 2.2, Theorem 3.1, and Corollary 3.2 of Chapter II can be generalized to a holomorphic map $f : L' \to L$. But other results of § 3 of Chapter II do not seem to generalize so easily.

Let E (resp. E') be a Hermitian vector bundle over a complex manifold M (resp. M'). From E (resp. E') we construct a Hermitian line bundle $L = \wedge^r E$ (resp. $L' = \wedge^{r'} E'$), where r (resp. r') is the fiber-dimension of E (resp. E'). Every holomorphic homomorphism $f : E' \to E$ induces a holomorphic homomorphism $f : L' \to L$ in a natural manner. Applying the results above to $f : L' \to L$ we obtain results on degeneracy of the mapping $f : E' \to E$ under suitable conditions on the Ricci tensors of E and E'. Appropriate conditions on the curvatures of E and E' could quite conceivably yield finer results if (1) of Theorem 1.1 is used. It would be interesting to find conditions which would guarantee that $f : E' \to E$ is norm-decreasing when f is suitably normalized.

IV

Invariant Distances on Complex Manifolds

1. An Invariant Pseudodistance

Let D denote the open unit disk in the complex plane $\mathbf{C}$ and let ϱ be the distance function on D defined by the Poincaré–Bergman metric of D.

Let M be a complex manifold. We define a pseudodistance d_M on M as follows. Given two points p, $q \in M$, we choose points $p = p_0, p_1, \ldots,$ $p_{k-1}, p_k = q$ of M, points $a_1, \ldots, a_k, b_1, \ldots, b_k$ of D, and holomorphic mappings $f_1, \ldots, f_k$ of D into M such that $f_i(a_i) = p_{i-1}$ and $f_i(b_i) = p_i$ for $i = 1, \ldots, k$. For each choice of points and mappings thus made, we consider the number

$$\varrho(a_1, b_1) + \cdots + \varrho(a_k, b_k).$$

Let $d_M(p, q)$ be the infimum of the numbers obtained in this manner for all possible choices. It is an easy matter to verify that $d_M : M \times M \to \mathbf{R}$ is continuous and satisfies the axioms for pseudodistance:

$$d_M(p, q) \geqq 0, \quad d_M(p, q) = d_M(q, p), \quad d_M(p, q) + d_M(q, r) \geqq d_M(p, r).$$

The most important property of d_M is given by the following proposition, the proof of which is trivial.

PROPOSITION 1.1. *Let M and N be two complex manifolds and let* $f : M \to N$ *be a holomorphic mapping. Then*

$$d_M(p, q) \geqq d_N(f(p), f(q)) \qquad \textit{for} \quad p, q \in M.$$

COROLLARY 1.2. *Every biholomorphic mapping* $f : M \to N$ *is an isometry, i.e.,*

$$d_M(p, q) = d_N(f(p), f(q)) \qquad for \quad p, q \in M.$$

The pseudodistance d_M may be considered as a generalization of the Poincaré–Bergman metric for a unit disk. We have

PROPOSITION 1.3. *For the open unit disk D in* **C**, d_D *coincides with the distance ϱ defined by the Poincaré–Bergman metric.*

Proof. By the Schwarz lemma (cf. Theorems 1.2 and 2.1 of Chapter I), every holomorphic mapping $f : D \to D$ is distance-decreasing with respect to ϱ. From the very definition of d_D we have

$$d_D(p, q) \geqq \varrho(p, q) \qquad for \quad p, q \in D.$$

Considering the identity transformation of D, we obtain the inequality $d_D(p, q) \leqq \varrho(p, q)$.   QED.

The following proposition says that d_M is the largest pseudodistance on M for which $f : D \to M$ is distance-decreasing.

PROPOSITION 1.4. *Let M be a complex manifold and d' any pseudodistance on M such that*

$$d'(f(a), f(b)) \leqq \varrho(a, b) \qquad for \quad a, b \in D$$

for every holomorphic mapping $f : D \to M$. Then

$$d_M(p, q) \geqq d'(p, q) \qquad for \quad p, q \in M.$$

Proof. Let $p_0, \ldots, p_k, a_1, \ldots, a_k, b_1, \ldots, b_k, f_1, \ldots, f_k$ be as in the definition of d_M. Then

$$d'(p, q) = \sum_{i=1}^{k} d'(p_{i-1}, p_i) = \sum_{i=1}^{k} d'(f_i(a_i), f_i(b_i))$$

$$\leqq \sum_{i=1}^{k} \varrho(a_i, b_i).$$

Hence,

$$d'(p, q) \leqq \inf \sum_{i=1}^{k} \varrho(a_i, b_i) = d_M(p, q).$$

QED.

PROPOSITION 1.5.   *Let M and M' be two complex manifolds. Then*

$$d_M(p, q) + d_{M'}(p', q') \geq d_{M \times M'}((p, p'), (q, q'))$$
$$\geq \mathrm{Max}[d_M(p, q), d_{M'}(p', q')]$$

$$for \quad p, q \in M \quad and \quad p', q' \in M'.$$

Proof.   We have

$$d_M(p, q) + d_{M'}(p', q') \geq d_{M \times M'}((p, p'), (q, p')) + d_{M \times M'}((q, p'), (q, q'))$$
$$\geq d_{M \times M'}((p, p'), (q, q')),$$

where the first inequality follows from the fact that the mappings $f : M \to M \times M'$ and $f' : M' \to M \times M'$ defined by $f(x) = (x, p')$ and $f'(x') = (q, x')$ are distance-decreasing and the second inequality is a consequence of the triangular axiom. The inequality $d_{M \times M'}((p, p'), (q, q')) \geq \mathrm{Max}[d_M(p, q), d_{M'}(p', q')]$ follows from the fact that the projections $M \times M' \to M$ and $M \times M' \to M'$ are both distance-decreasing.   QED.

Example 1.   *If D is the open unit disk in $\mathbf{C}$, then*

$$d_{D \times D}((p, p'), (q, q')) = \mathrm{Max}[d_D(p, q), d_D(p', q')] \quad for \quad p, q, p', q' \in D.$$

To prove this assertion we may assume (because of the homogeneity of D) that $p = p' = 0$ and $d_D(0, q) \geq d_D(0, q')$, i.e., $|q| \geq |q'|$. Consider the holomorphic mapping $f : D \to D \times D$ defined by $f(z) = (z, (q'/q)z)$. Since f is distance-decreasing, we have

$$d_{D \times D}((0, 0), (q, q')) = d_{D \times D}(f(0), f(q)) \leq d_D(0, q).$$

More generally, *if D^k denotes the k-dimensional polydisk $D \times \cdots \times D$, then*

$$d_{D^k}((p_1, \ldots, p_k), (q_1, \ldots, q_k)) = \mathrm{Max}[d_D(p_i, q_i); \ i = 1, \ldots, k].$$

This shows that d_{D^k} does not coincides with the distance defined by the Bergman metric of D^k unless $k = 1$.

As we see from the definition of the pseudodistance, d_M is defined in a manner similar to the distance function on a Riemannian manifold. It is therefore quite natural to expect the following result:

PROPOSITION 1.6.   *Let M be a complex manifold and $\tilde{M}$ a covering manifold of M with covering projection $\pi : \tilde{M} \rightarrow M$. Let $p, q \in M$ and $\tilde{p}, \tilde{q} \in \tilde{M}$ such that $\pi(\tilde{p}) = p$ and $\pi(\tilde{q}) = q$. Then*

$$d_M(p, q) = \inf_{\tilde{q}} d_{\tilde{M}}(\tilde{p}, \tilde{q}),$$

where the infimum is taken over all $\tilde{q} \in \tilde{M}$ such that $q = \pi(\tilde{q})$.

Proof.   Since $\pi : \tilde{M} \rightarrow M$ is distance-decreasing, we have

$$d_M(p, q) \leq \inf_{\tilde{q}} d_{\tilde{M}}(\tilde{p}, \tilde{q}).$$

Assuming the strict inequality, let

$$d_M(p, q) + \varepsilon < \inf_{\tilde{q}} d_{\tilde{M}}(\tilde{p}, \tilde{q}),$$

where ε is a positive number. By the very definition of d_M there exist points $a_1, \ldots, a_k, b_1, \ldots, b_k$ of the unit disk D and holomorphic mappings $f_1, \ldots, f_k$ of D into M such that

$$p = f_1(a_1), \quad f_1(b_1) = f_2(a_2), \quad \cdots \quad f_{k-1}(b_{k-1}) = f_k(a_k), \quad f_k(b_k) = q$$

and

$$d_M(p, q) + \varepsilon > \sum_{i=1}^{k} \varrho(a_i, b_i).$$

Then we can lift $f_1, \ldots, f_k$ to holomorphic mappings $\tilde{f}_1, \ldots, \tilde{f}_k$ of D into $\tilde{M}$ in such a way that

$$\tilde{p} = \tilde{f}_1(a_1),$$
$$\tilde{f}_i(b_i) = \tilde{f}_{i+1}(a_{i+1}) \qquad \text{for} \quad i = 1, \ldots, k-1,$$
$$\pi \circ \tilde{f}_i = f_i \qquad \text{for} \quad i = 1, \ldots, k.$$

If we set $\tilde{q} = \tilde{f}_k(b_k)$, then $\pi(\tilde{q}) = q$ and $d_{\tilde{M}}(\tilde{p}, \tilde{q}) \leq \sum_{i=1}^{k} \varrho(a_i, b_i)$. Hence, $d_{\tilde{M}}(\tilde{p}, \tilde{q}) < d_M(p, q) + \varepsilon$, which contradicts our assumption.   QED.

Remark.   It is not clear whether the following is true:

$$d_M(p, q) = \min_{\tilde{q}} d_{\tilde{M}}(\tilde{p}, \tilde{q}),$$

i.e., whether $\inf_{\tilde{q}} d_{\tilde{M}}(\tilde{p}, \tilde{q})$ is really attained by some $\tilde{q}$.

Example 2.   Let M be a Riemann surface whose universal covering is a unit disk D, e.g., a compact Riemann surface of genus ≥ 2. From Propositions 1.3 and 1.6 we see that d_M is the distance defined by a Kaehler metric of M with constant curvature <0.

Example 3.   It can be seen easily from the definition of the pseudo-distance d_M that *for a complex Euclidean space* $\mathbf{C}^n$ *the pseudodistance* $d_{\mathbf{C}^n}$ *is trivial, i.e.,* $d_{\mathbf{C}^n}(p, q) = 0$ *for all* $p, q \in \mathbf{C}^n$. More generally, *if* M *is a complex manifold on which a complex Lie group* G *acts transitively, then* d_M *is trivial.* For each point $p \in M$, let U be a neighborhood of p such that every point $q \in U$ lies on the orbit of a complex 1-parameter subgroup of G through p. Hence, for every point $q \in U$ there is a holomorphic mapping $f : \mathbf{C} \to M$ whose image contains both p and q. Since f is distance-decreasing and $d_{\mathbf{C}}$ is trivial, we have $d_M(p, q) = 0$. To prove that d_M is trivial for any pair of points on M, we connect the pair by a chain of small open sets U and apply the triangular axiom. In particular, *for a compact homogeneous complex manifold* M *or for* $M = \mathbf{C} - \{0\}$, d_M *is trivial. If* M *is a compact Riemann surface of genus* 0 *or* 1, *then* d_M *is trivial.*

Example 4.   Let $M = \{z \in \mathbf{C}^n; r < |z| < R\}$ and $B = \{z \in \mathbf{C}^n; |z| < R\}$. If $n = 1$, the universal covering space of M is the upper half-plane or equivalently the unit disk. In § 6 of Chapter I we showed how to construct a metric of constant negative curvature $ds_M{}^2$ from the Poincaré–Bergman metric of the upper half-plane. From Proposition 1.6 it follows that the invariant distance d_M coincides with the distance defined by $ds_M{}^2$. For $n > 1$, I do not know what d_M looks like, although I know that d_M is not complete, since M is not holomorphically complete (see § 4 of this chapter).

2. *Carathéodory Distance*

Let D be the open unit disk in $\mathbf{C}$ with the Poincaré–Bergman distance ϱ. Let M be a complex manifold. The *Carathéodory pseudodistance* c_M of M is defined by

$$c_M(p, q) = \sup_f \varrho(f(p), f(q)) \qquad \text{for} \quad p, q \in M,$$

where the supremum is taken with respect to the family of holomorphic mappings $f : M \to D$. We prove that the pseudodistance d_M defined in

§ 1 is greater than or equal to c_M. That will in particular imply that $c_M(p, q)$ is finite.

PROPOSITION 2.1.   *For any complex manifold M, we have*

$$d_M(p, q) \geqq c_M(p, q) \qquad for \quad p, q \in M.$$

Proof.   As in the definition of $d_M(p, q)$, choose points $p = p_0, p_1, \ldots,$ $p_{k-1}, p_k = q$ of M and points $a_1, \ldots, a_k, b_1, \ldots, b_k$ of D and also mappings $f_1, \ldots, f_k$ of D into M such that $f_i(a_i) = p_{i-1}$ and $f_i(b_i) = p_i$. Let f be a holomorphic mapping of M into D. Then

$$\sum_{i=1}^{k} \varrho(a_i, b_i) \geqq \sum_{i=1}^{k} \varrho(f \circ f_i(a_i), f \circ f_i(b_i))$$
$$\geqq \varrho(f \circ f_1(a_1), f \circ f_k(b_k))$$
$$= \varrho(f(p), f(q)),$$

where the first inequality follows from the Schwarz lemma and the second inequality is a consequence of the triangular axiom. Hence,

$$d_M(p, q) = \inf \sum_{i=1}^{k} \varrho(a_i, b_i) \geqq \sup \varrho(f(p), f(q)) = c_M(p, q).$$

QED.

It is an easy matter to verify that $c_M : M \times M \to R$ is continuous and satisfies the axioms for pseudodistance:

$$c_M(p, q) \geqq 0, \qquad c_M(p, q) = c_M(q, p), \qquad c_M(p, q) + c_M(q, r) \geqq c_M(p, r).$$

The Carathéodory pseudodistance c_M shares many properties with d_M.

PROPOSITION 2.2.   *Let M and N be two complex manifolds and let* $f : M \to N$ *be a holomorphic mapping. Then*

$$c_M(p, q) \geqq c_N(f(p), f(q)) \qquad for \quad p, q \in M.$$

The proof is trivial.

COROLLARY 2.3.   *Every biholomorphic mapping* $f : M \to N$ *is an isometry, i.e.,*

$$c_M(p, q) = c_N(f(p), f(q)) \qquad for \quad p, q \in M.$$

The Carathéodory pseudodistance may be also considered as a generalization of the Poincaré–Bergman metric for D.

PROPOSITION 2.4. *For the open unit disk D in $\mathbf{C}$, c_D coincides with the distance ϱ defined by the Poincaré–Bergman metric.*

Proof. Using the Schwarz lemma for a holomorphic mapping $f : D \to D$ we obtain

$$\varrho(p, q) \geqq c_D(p, q) \qquad \text{for} \quad p, q \in D$$

from the very definition of c_D. Considering the identity transformation of D, we obtain the inequality $\varrho(p, q) \leqq c_D(p, q)$.   QED.

The following proposition says that c_M is the smallest pseudodistance on M for which $f : M \to D$ is distance-decreasing.

PROPOSITION 2.5. *Let M be a complex manifold and c' any pseudodistance on M such that*

$$c'(p, q) \geqq \varrho(f(p), f(q)) \qquad p, q \in M$$

for every holomorphic mapping $f : M \to D$. Then

$$c_M(p, q) \leqq c'(p, q) \qquad \text{for} \quad p, q \in M$$

The proof is trivial.

The following proposition can be proved in the same way as Proposition 1.5.

PROPOSITION 2.6. *Let M and M' be two complex manifolds. Then*

$$c_M(p, q) + c_{M'}(p', q') \geqq c_{M \times M'}((p, p'), (q, q'))$$
$$\geqq \text{Max}[c_M(p, q), c_{M'}(p', q')]$$
$$\text{for} \quad p, q \in M \quad \text{and} \quad p', q' \in M'.$$

Example 1. *If D^k denotes the k-dimensional polydisk $D \times \cdots \times D$, then*

$$c_{D^k}((p_1, \ldots, p_k), (q_1, \ldots, q_k)) = \text{Max}[c_D(p_i, q_i); \ i = 1, \ldots, k]$$
$$\text{for} \quad p_i, q_i \in D.$$

The proof is similar to the one in Example 1 of § 1. Hence,

$$d_{D^k} = c_{D^k}.$$

Example 2.   Let $M = G/K$ be a symmetric bounded domain of rank l in $\mathbf{C}^n$ (in its natural realization). For a suitable subspace $\mathbf{C}^l$ of $\mathbf{C}^n$, $M \cap \mathbf{C}^l$ is a polydisk $D^l = D \times \cdots \times D$ and M can be written as a union of polydisks $k(D^l)$, $k \in K$, where k is considered as a unitary transformation of $\mathbf{C}^n$. In particular, we have

$$D^l \subset M \subset D^n.$$

Since the injections $D^l \to M$ and $M \to D^n$ are distance-decreasing, we have $c_{D^l} \geq c_M \geq c_{D^n}$ on D^l. From Example 1 we see that $c_{D^l} = c_{D^n}$ on D^l. Hence,

$$c_{D^l} = c_M \qquad \text{on} \quad D^l.$$

Using the result in Example 1 of § 1, we obtain a similar result for d_{D^l} and d_M. Given any two points $p, q \in M$, there is an element of G which sends p and q into D^l. We may conclude therefore that, *for a symmetric bounded domain M, the two distances c_M and d_M coincide.*

Example 3.   Since $c_M \leq d_M$ in general, we see from Example 3 of § 1 that *if M is a complex manifold on which a complex Lie group G acts transitively, then c_M is trivial.*

Example 4.   *If M is a compact complex manifold, then c_M is trivial.* This is clear, since every holomorphic function on M must be a constant function. This shows that there is no analogue of Proposition 1.6 for c_M.

Example 5.   *Let $M = \mathbf{C} - A$, where A is a finite set of points, or more generally, let $M = \mathbf{C}^n - A$, where A is an analytic subset of dimension $\leq n - 1$. Then c_M is trivial.* A holomorphic mapping $f : M \to D$ may be considered as a bounded holomorphic function on M. By the so-called Riemann extension theorem, f can be extended to a holomorphic mapping of $\mathbf{C}^n$ into D. By Liouville's theorem (or by Example 3), f must be a constant function and hence c_M is trivial.

Example 6.   *Let $M = \{z \in \mathbf{C}^n; \ r < |z| < R\}$ and $B = \{z \in \mathbf{C}^n; \ |z| < R\}$. If $n \geq 2$, then $c_M = c_B | M$.* This follows from the fact that the envelop of holomorphy of M is B, so that every holomorphic mapping from M into the unit disk D can be extended to a holomorphic mapping from B into D. On the other hand, I do not know what the Carathéodory distance c_M looks like if $n = 1$. This is in contrast to the situation in Example 4 of § 1, where d_M is known for $n = 1$ but not for $n \geq 2$.

The Carathéodory distance was introduced in Carathéodory [1, 2]. For recent results on the Carathéodory distance, see Reiffen [1, 2].

3. *Completeness with Respect to the Carathéodory Distance*

Let M be a complex manifold. Throughout this section we are interested in the case where the Carathéodory pseudodistance c_M is a distance, i.e., the case where the family of holomorphic mappings $f : M \to D$ separates the points of M. Since every bounded holomorphic function on M, multiplied by a suitable constant, yields a holomorphic mapping of M into the unit disk D, the necessary and sufficient condition that c_M be a distance is the following: *For any two distinct points p and q of M, there is a bounded holomorphic function f on M such that $f(p) \neq f(q)$.* It is very close to assuming that M is a bounded domain in a Stein manifold. In particular, all compact complex manifolds are excluded.

In general, we say that a metric space M is *complete* if, for each point p of M and each positive number r, the closed ball of radius r around p is a compact subset of M. If M is complete in this sense, then every Cauchy sequence of M converges. The converse is true for a Riemannian metric, but not in general.

PROPOSITION 3.1. *If M and M' are complex manifolds with complete Carathéodory distance, so is $M \times M'$.*

Proof. This is immediate from Proposition 2.6.   QED.

PROPOSITION 3.2. *A closed complex submanifold M' of a complex manifold M with complete Carathéodory distance is also complete with respect to its Carathéodory distance.*

Proof. This is immediate from the fact that the injection $M' \to M$ is distance-decreasing.   QED.

PROPOSITION 3.3. *Let M and M_i, $i \in I$, be complex submanifolds of a complex manifold N such that $M = \cap_i M_i$. If each M_i is complete with respect to its Carathéodory distance, so is M.*

Proof. Since each injection $M \to M_i$ is distance-decreasing, the proposition follows from the following trivial lemma.

LEMMA. *Let M and M_i, $i \in I$, be subsets of a topological space N such* *that $M = \cap_i M_i$. Let d and d_i be distances on M and M_i such that $d(p, q)$ $\geqq d_i(p, q)$ for $p, q \in M$. If each M_i is complete with respect to d_i, then M is complete with respect to d.*   QED.

We shall now give a large class of bounded domains which are complete with respect to their Carathéodory distances. Let G be a domain in $\mathbf{C}^n$ and $f_1, \ldots, f_k$ holomorphic functions defined in G. Let P be a connected component of the open subset of G defined by

$$|f_1(z)| < 1, \ldots, |f_k(z)| < 1.$$

Assume that the closure of P is compact and is contained in G. Then P is called an *analytic polyhedron*.

THEOREM 3.4.   *An analytic polyhedron P is complete with respect to its Carathéodory distance c_P.*

Proof.   Let F be the set of holomorphic mappings of P into the unit disk D. Let o be a point of P. Given a positive number a, choose a positive number b, $0 < b < 1$, such that

$$\{z \in D;\ \varrho(f_i(o), z) \leqq a \quad \text{for} \quad i = 1, \ldots, k\} \subset \{z \in D;\ |z| \leqq b\}.$$

Then

$$
\begin{aligned}
\{p \in P;\ c_P(o, p) \leqq a\} &= \{p \in P;\ \varrho(f(o), f(p)) \leqq a \text{ for } f \in F\} \\
&\subset \{p \in P;\ \varrho(f_i(o), f_i(p)) \leqq a \text{ for } i = 1, \ldots, k\} \\
&\subset \{p \in P;\ |f_i(p)| \leqq b \text{ for } i = 1, \ldots, k\}.
\end{aligned}
$$

Since $\{p \in P;\ |f_i(p)| \leqq b$ for $i = 1, \ldots, k\}$ is compact, $\{p \in P;\ c_P(o, p) \leqq a\}$ is compact.   QED.

In the definition of an analytic polyhedron, we used only a finitely many holomorphic functions $f_1, \ldots, f_k$. If we take an arbitrary family of holomorphic functions f_α, $\alpha \in A$, on G and assume that the set defined by $|f_\alpha(z)| < 1$, $\alpha \in A$, is open, then we call each connected component of this open subset of G a *generalized analytic polyhedron*. The proof of Theorem 3.4 is valid for the following:

THEOREM 3.5.   *A generalized analytic polyhedron P is complete with respect to its Carathéodory distance c_P.*

Let h_j be real analytic functions on G of the form

$$h_j = \sum_{m=1}^{\infty} |f_{jm}|^2, \qquad j = 1, \ldots, k,$$

where each f_{jm} is holomorphic in G. Let S be the set of sequences $\{a_m\}$ of complex numbers such that $\sum_{m=1}^{\infty} |a_m|^2 = 1$. Then

$$\{z \in G; \; |h_j(z)| < 1\} = \bigcap_{\{a_m\} \in S} \left\{ z \in G; \; \left| \sum_{m=1}^{\infty} a_m f_{jm}(z) \right| < 1 \right\}.$$

This shows that a connected component of the open subset of G defined by

$$\{z \in G; \; |h_j(z)| < 1 \text{ for } j = 1, \ldots, k\}$$

is a generalized analytic polyhedron. To such a manifold, Theorem 3.5 is therefore applicable.

So far, we have given sufficient conditions for a manifold to be complete with respect to its Carathéodory distance. We shall give now some neccessary conditions.

Horstmann [1] has shown that a domain in $\mathbf{C}^n$ which is complete with respect to its Carathéodory distance is a domain of holomorphy. We shall give a generalization of Horstmann's result. Given a family F of holomorphic functions on a complex manifold M and a subset K of M, we set

$$\hat{K}_F = \{p \in M; \; |f(p)| \leqq \sup |f(K)| \text{ for all } f \in F\},$$

where $\sup |f(K)|$ denotes the supremum of $|f(q)|$ for $q \in K$. Then $\hat{K}_F$ is a closed subset of M containing K and is called the *convex hull* of K with respect to F. The convex hull of $\hat{K}_F$ with respect to F coincides with $\hat{K}_F$ itself. If $\hat{K}_F$ is compact for every compact subset K of M, then M is said to be *convex with respect to F*. If $F' \subset F$, then $\hat{K}_{F'} \supset \hat{K}_F$. Hence, if M is convex with respect to F' and if $F' \subset F$, then M is convex with respect to F. If M is convex with respect to the family of all holomorphic functions on M, then M is said to be *holomorphically convex*.

THEOREM 3.6.   *Let M be a complex manifold with Carathéodory distance. Fix a point o of M and let F be the set of bounded holomorphic functions f on M such that $f(o) = 0$. If M is complete with respect to c_M, then M is convex with respect to F and hence is holomorphically convex.*

Proof. Let a be a positive number and B the closed ball of radius a around $o \in M$, i.e.,

$$B = \{p \in M; \ c_M(o, p) \leqq a\}.$$

Since M is complete, B is compact. Since every compact subset K of M is contained in B for a sufficiently large a, it suffices to prove that $\hat{B}_F$ is compact. We shall actually show that $\hat{B}_F = B$.

$$
\begin{aligned}
\hat{B}_F &= \{p \in M; \ |f(p)| \leqq \sup |f(B)| \ \text{for} \ f \in F\} \\
&= \{p \in M; \ \varrho(0, f(p)) \leqq \sup_{q \in B} \varrho(0, f(q)) \ \text{for} \ f \in F\} \\
&\subset \{p \in M; \ \varrho(0, f(p)) \leqq \sup_{q \in B} c_M(o, q)\} \\
&= \{p \in M; \ \varrho(0, f(p)) \leqq a\} \\
&= B.
\end{aligned}
$$

Since $\hat{B}_F$ contains B, we conclude $\hat{B}_F = B$.   QED.

If f is a bounded holomorphic function on M, then, for a suitable positive constant c, cf is a holomorphic mapping of M into the open unit disk D. This means that, in Theorem 3.6, we may choose F to be the family of holomorphic mappings $f : M \to D$ such that $f(o) = 0$.

It is not clear if the converse to Theorem 3.6 holds. As we shall see shortly, M need not be complete with respect to c_M even if M is holomorphically convex. It seems, however, very reasonable to expect that M is complete with respect to c_M if it is convex with respect to the family of bounded holomorphic functions.

Let M be a complex manifold of complex dimension n and A an analytic subset of dimension $\leqq n - 1$. Since every bounded holomorphic functions on $M - A$ can be extended to a bounded holomorphic function on M by Riemann's extension theorem, it follows that $M - A$ cannot be convex with respect to the family of bounded holomorphic functions if A is nonempty and that c_M coincides with c_{M-A} on $M - A$. In this way we obtain many examples of holomorphically convex manifolds which are not convex with respect to the family of bounded holomorphic functions and not complete with respect to their Carathéodory distances. The punctured disk $D - \{0\}$ is the simplest example. The question of completeness with respect to the Carathéodory distance may be quite possibly related to that of completeness with respect to the Bergman metric (see Bremermann [1] and Kobayashi [1] on the latter question).

Since D is the completion of $D - \{0\}$ with respect to the Carathéodory distance $c_{D-\{0\}}$ and every holomorphic mapping $f : D - \{0\} \to M$ is distance-decreasing with respect to $c_{D-\{0\}}$ and c_M, we have

PROPOSITION 3.7. *If M is a complex manifold with complete Carathéodory distance, then every holomorphic mapping f from the punctured disk $D - \{0\}$ into M can be extended to a holomorphic mapping of D into M.*

4. *Hyperbolic Manifolds*

Let M be a complex manifold and d_M the pseudodistance defined in § 1. If d_M is a distance, i.e., $d_M(p, q) > 0$ for $p \neq q$, then M is called a *hyperbolic manifold*. A hyperbolic manifold M is said to be *complete* if it is complete with respect to d_M. As in § 3, M is complete with respect to d_M if, for each point p of M and each positive number r, the closed ball of radius r around p is a compact subset of M. As we shall see later, for a hyperbolic manifold M, this definition is equivalent to the usual definition in terms of Cauchy sequences.

The proofs of the following three propositions are almost identical to those of Propositions 3.1, 3.2, and 3.3.

PROPOSITION 4.1. *If M and M' are (complete) hyperbolic manifolds, so is $M \times M'$.*

PROPOSITION 4.2. *A (closed) complex submanifold M' of a (complete) hyperbolic manifold M is a (complete) hyperbolic manifold.*

PROPOSITION 4.3. *Let M and M_i, $i \in I$, be closed complex submanifolds of a complex manifold N such that $M = \cap_i M_i$. If each M_i is a (complete) hyperbolic manifold, so is M.*

From Proposition 2.1 we obtain

PROPOSITION 4.4. *A complex manifold M is (complete) hyperbolic if its Carathéodory pseudodistance c_M is a (complete) distance.*

COROLLARY 4.5. *Every bounded domain in $\mathbf{C}^n$ is hyperbolic. Every generalized analytic polyhedron is complete hyperbolic.*

The second statement follows from Theorem 3.5.

The following proposition is immediate from Proposition 1.4.

PROPOSITION 4.6. *If a complex manifold M admits a (complete) distance d' for which every holomorphic mapping f of the open unit disk D into M is distance-decreasing, i.e., $d'(f(a), f(b)) \leq \varrho(a, b)$ for $a, b \in D$, then it is (complete) hyperbolic.*

The results so far have their counterparts for Carathéodory distance. We shall now give results which are proper to the invariant distance d_M.

THEOREM 4.7. *Let M be a complex manifold and $\tilde{M}$ a covering manifold of M. Then $\tilde{M}$ is (complete) hyperbolic if and only if M is (complete) hyperbolic.*

Proof. Assume that $\tilde{M}$ is hyperbolic. Let $p, q \in M$ and $d_M(p, q) = 0$. Let $\tilde{p}$ be a point of $\tilde{M}$ such that $\pi(\tilde{p}) = p$, where $\pi : \tilde{M} \to M$ is the covering projection. By Proposition 1.6, there exists a sequence of points $\tilde{q}_1, \ldots, \tilde{q}_i, \ldots$ of $\tilde{M}$ such that $\pi(\tilde{q}_i) = q$ and $\lim_i d_{\tilde{M}}(\tilde{p}, \tilde{q}_i) = 0$. Then the sequence $\{\tilde{q}_i\}$ converges to $\tilde{p}$. Hence $\pi(\tilde{q}_i)$ converges to p. Since $\pi(\tilde{q}_i) = q$, we obtain $p = q$. This proves that M is hyperbolic.

Assume that $\tilde{M}$ is complete hyperbolic. Let $\tilde{B}_r$ be the closed ball of radius r around $\tilde{p} \in \tilde{M}$, i.e., $\tilde{B}_r = \{\tilde{q} \in \tilde{M}; d_{\tilde{M}}(\tilde{p}, \tilde{q}) \leq r\}$. Similarly, let B_r be the closed ball of radius r around $p = \pi(\tilde{p}) \in M$. By Proposition 1.6, we have

$$B_r \subset \pi(\tilde{B}_{r+\delta}) \qquad \text{for} \quad \delta > 0.$$

Since $\tilde{B}_{r+\delta}$ is compact by assumption and B_r is closed, B_r must be compact. Hence, M is complete.

Assume that M is hyperbolic. Let $\tilde{p}, \tilde{q} \in \tilde{M}$ and $d_{\tilde{M}}(\tilde{p}, \tilde{q}) = 0$. Since the projection π is distance-decreasing, we have $d_M(\pi(\tilde{p}), \pi(\tilde{q})) = 0$, which implies $\pi(\tilde{p}) = \pi(\tilde{q})$. Let $\tilde{U}$ be a neighborhood of $\tilde{p}$ in $\tilde{M}$ such that $\pi : \tilde{U} \to \pi(\tilde{U})$ is a diffeomorphism and $\pi(\tilde{U})$ is an ε-neighborhood of $\pi(\tilde{p})$ with respect to d_M. In particular, $\tilde{U}$ does not contain $\tilde{q}$ unless $\tilde{p} = \tilde{q}$. Since $d_{\tilde{M}}(\tilde{p}, \tilde{q}) = 0$ by assumption, there exist points $a_1, \ldots, a_k$, $b_1, \ldots, b_k$ of D and holomorphic mappings $f_1, \ldots, f_k$ of D into $\tilde{M}$ such that $\tilde{p} = f_1(a_1)$, $f_i(b_i) = f_{i+1}(a_{i+1})$ for $i = 1, \ldots, k-1$ and $f_k(b_k) = \tilde{q}$ and that $\sum_{i=1}^k \varrho(a_i, b_i) < \varepsilon$. Let $\widehat{a_i b_i}$ denote the geodesic arc from a_i to b_i in D. Joining the curves $f_1(\widehat{a_1 b_1}), \ldots, f_k(\widehat{a_k b_k})$ in $\tilde{M}$, we obtain a curve from $\tilde{p}$ to $\tilde{q}$ in $\tilde{M}$, which will be denoted by $\tilde{C}$. Since $\pi \circ f_1, \ldots, \pi \circ f_k$ are distance-decreasing mappings of D into M and $\widehat{a_1 b_1}, \ldots, \widehat{a_k b_k}$ are geodesics in D, every point of the curve $\pi(\tilde{C})$ remains in the ε-neighborhood $\pi(\tilde{U})$ of $\pi(\tilde{p})$. Hence, the endpoint $\tilde{q}$ must coincide with $\tilde{p}$.

Assume that M is complete hyperbolic. We shall prove here that $\tilde{M}$ is complete in the sense that every Cauchy sequence with respect to $d_{\tilde{M}}$ is convergent. In the following section, we shall prove that the completeness in this usual sense implies the completeness defined at the beginning of this section. Let $\{\tilde{p}_i\}$ be a Cauchy sequence in $\tilde{M}$. Since the projection π is distance-decreasing, $\{\pi(\tilde{p}_i)\}$ is a Cauchy sequence in M and hence converges to a point $p \in M$. Let ε be a positive number and U the 2ε-neighborhood of p in M. Taking ε small we may assume that π induces a homeomorphism of each connected component of $\pi^{-1}(U)$ onto U. Let N be a large integer such that $\pi(\tilde{p}_i)$ is within the ε-neighborhood of p for $i > N$. Then every point outside U is at least ε away from $\pi(\tilde{p}_i)$. Let $\tilde{U}_i$ be the connected component of $\pi^{-1}(U)$ containing $\tilde{p}_i$. We shall show that the ε-neighborhood of $\tilde{p}_i$ lies in $\tilde{U}_i$ for $i > N$. Let $\tilde{q}$ be a point of $\tilde{M}$ with $d_{\tilde{M}}(\tilde{p}_i, \tilde{q}) < \varepsilon$. We choose points $a_1, \ldots, a_k, b_1, \ldots, b_k$ of D and holomorphic mappings $f_1, \ldots, f_k$ of D into $\tilde{M}$ in the usual manner so that $\sum_{j=1}^{k} \varrho(a_j, b_j) < \varepsilon$. Denoting by $\widehat{a_j b_j}$ the geodesic arc from a_j to b_j, let $\tilde{C}$ be the curve from $\tilde{p}_i$ to $\tilde{q}$ obtained by joining $f_1(\widehat{a_1 b_1})$, $\ldots, f_k(\widehat{a_k b_k})$ in M. Let $C = \pi(\tilde{C})$. From the construction of C, it is clear that C is contained in the ε-neighborhood of $\pi(p_i)$ and hence in U. It follows that $\tilde{C}$ lies in $\tilde{U}_i$. Let $\tilde{p}$ be the point of $\tilde{U}_i$ defined by $p = \pi(\tilde{p})$. Then $\{\tilde{p}_j\}$ converges to $\tilde{p}$.   QED.

A complex manifold $\tilde{M}$ is called a *spread* (*domaine etalé* in French) over a complex manifold M with projection π if every point $\tilde{p} \in \tilde{M}$ has a neighborhood $\tilde{U}$ such that π is a holomorphic diffeomorphism of $\tilde{U}$ onto the open set $\pi(\tilde{U})$ of M. This is a concept more general than that of covering manifold. From the proof of Theorem 4.7 above, we obtain the following:

PROPOSITION 4.8.   *A spread $\tilde{M}$ over a hyperbolic manifold M is hyperbolic.*

A similar reasoning yields also the following:

PROPOSITION 4.9.   *If a complex manifold M' is holomorphically immersed in a hyperbolic manifold M, then M' is also hyperbolic.*

In § 3 we saw that the punctured disk $D - \{0\}$ is not complete with respect to its Carathéodory distance. But it is complete hyperbolic. In fact, the half-plane $\{z = x + iy \in \mathbf{C}; \ y > 0\}$ is the universal covering space of $D - \{0\}$, the projection being given by $z \to e^{iz}$. We may also consider the unit disk D as the universal covering space of $D - \{0\}$. Our assertion follows from Theorem 4.7. More generally, we have

THEOREM 4.10.   *Let M be a complete hyperbolic manifold and f a bounded holomorphic function on M. Then the open submanifold $M' = \{p \in M; \ f(p) \neq 0\}$ of M is also complete hyperbolic.*

Proof.   Multiplying f by a suitable constant, we may assume that f is a holomorphic mapping of M into the open unit disk D. We denote by D^* the punctured disk $D - \{0\}$. Let o be a point of M' and let a and b be positive numbers. Since D^* is complete hyperbolic, for a given positive number a we can choose a small positive number b such that

$$\{z \in D; \ |z| \geq b\} \supset \{z \in D^*; \ d_{D^*}(f(o), z) \leq a\}.$$

We set

$$A = \{p \in M; \ d_M(o, p) \leq a\}, \qquad A' = \{p \in M'; \ d_{M'}(o, p) \leq a\},$$
$$B = \{p \in M; \ |f(p)| \geq b\}, \qquad B' = \{p \in M'; \ |f(p)| \geq b\}.$$

Since $d_{M'}(o, p) \geq d_M(o, p)$ by Proposition 1.1, we have

$$A \supset A'.$$

Since b is positive and $M' = \{p \in M; \ f(p) \neq 0\}$, we have

$$B = B'.$$

Since $f : M' \to D^*$ is distance-decreasing, we have

$$A' \subset \{p \in M'; d_{D^*}(f(o), f(p)) \leq a\} \subset \{p \in M'; |f(p)| \geq b\} = B' = B.$$

Since A is compact subset of M by the completeness of M and B is closed in M, the intersection $A \cap B$ is compact. Since $B = B'$, it follows that $A \cap B$ is in M'. Since $A \cap B$ is a compact subset of M' and A' is closed in M', the intersection $A' \cap (A \cap B)$ is a compact subset of M'. Since both A and B contain A', it follows that A' coincides with $A' \cap (A \cap B)$ and hence is a compact subset of M', thus proving that M' is complete hyperbolic.   QED.

It should be observed that M' above can never be complete with respect to its Carathéodory distance $c_{M'}$ unless $M = M'$. In fact, $c_{M'} = c_M \mid M'$ (see § 3).

THEOREM 4.11.   *A (complete) Hermitian manifold M whose holomorphic sectional curvature is bounded above by a negative constant is (complete) hyperbolic.*

Proof.   Let $ds_D{}^2$ denote the Poincaré-Bergman metric on the open unit disk D. Let $ds_M{}^2$ be the Hermitian metric of M. If we multiply $ds_M{}^2$ by a suitable positive constant, we have by Theorem 2.1 of Chapter III the following inequality for every holomorphic mapping $f : D \to M$.

$$f^*(ds_M{}^2) \leqq ds_D{}^2.$$

If we denote by d' and ϱ the distance functions on M and D defined by $ds_M{}^2$ and $ds_D{}^2$ respectively, then this inequality implies that f is distance-decreasing with respect to d' and ϱ, i.e., $d'(f(a), f(b)) \leqq \varrho(a, b)$ for $a, b \in D$. Now the theorem follows from Proposition 4.6.   QED.

COROLLARY 4.12.   *The Gaussian plane minus two points $\mathbf{C} - \{a, b\}$ is a complete hyperbolic manifold.*

This follows from the result in § 3 of Chapter I and Theorem 4.11. Similarly, the following corollary follows from Theorem 5.1 of Chapter I.

COROLLARY 4.13.   *Every compact Riemann surface of genus $g \geqq 2$ is a compact hyperbolic manifold.*

As a consequence of and also as a generalization of Corollary 4.12 we have the following example (due to P. J. Kiernan).

Example 1.   In $P_2(\mathbf{C})$, let L_1, L_2, L_3, and L_4 be four (complex) lines in general position. Let $a = L_1 \cap L_2$ and $b = L_3 \cap L_4$. Let L_0 be the lines through a and b. Then $M = P_2(\mathbf{C}) - \cup_{i=0}^{4} L_i$ is complete hyperbolic. In fact, M is biholomorphic to the direct product of two copies of $\mathbf{C} - \{0, 1\}$. To see this, we consider L_0 as the line at infinity so that $P_2(\mathbf{C}) - L_0 = \mathbf{C}^2$. Then L_1 is parallel to L_2, since $L_1 \cap L_2 = a \in L_0$. Similarly, L_3 is parallel to L_4. But L_1 and L_3 are not parallel, since L_1, L_2, L_3, and L_4 are in general position in $P_2(\mathbf{C})$. It is now clear that $M = \mathbf{C}^2 - \cup_{i=1}^{4} L_i$ is affinely equivalent to the direct product of two copies of $\mathbf{C} - \{0, 1\}$. On the other hand, $P_2(\mathbf{C}) - \cup_{i=1}^{4} L_i$ is not hyperbolic, since it contains $L_0 - \{a, b\} = P_1(\mathbf{C}) - \{a, b\}$, which is not hyperbolic. Let p_j, $j = 1, \ldots, 4$, be four points in general position in $P_2(\mathbf{C})$. Connecting every pair of these points, we obtain six lines L_j, $j = 0, 1, \ldots, 5$. Since $P_2(C) - \cup_{j=0}^{5} L_j$ is contained in $M = P_2(C) - \cup_{i=0}^{4} L_j$, it is also hyperbolic. I do not know whether it is also complete. For higher-dimensional analogues of this example, see Kiernan [1].

The following result is also due to Kiernan [4].

THEOREM 4.14.   *Let E be a holomorphic fiber bundle over M with fibre F and projection π. Then E is (complete) hyperbolic if M and F are (complete) hyperbolic.*

Proof.   Assume that M and F are hyperbolic. Let $p, q \in E$. If $\pi(p) \neq \pi(q)$, then $d_E(p, q) \geq d_M(\pi(p), \pi(q)) > 0$. Assume $\pi(p) = \pi(q)$. Choose a neighborhood U of $\pi(p)$ in M such that $\pi^{-1}(U) = U \times F$. Let B_s be the ball in M centered at $\pi(p)$ and of radius s with respect to d_M. Denote by D_r the disk $\{z \in \mathbf{C}; \ |z| < r\}$. Choose $s > 0$ and $r > 0$ in such a way that $B_{2s} \subset U$ and $d_D(z, 0) < s$ for $z \in D_r$ (where $D = D_1$). Thus, if $f : D \to E$ is holomorphic and $f(0) \in \pi^{-1}(B_s)$, then $f(D_r) \subset U \times F$. Choose $c > 0$ such that $d_D(0, a) \geq c \, d_{D_r}(0, a)$ for all $a \in D_{r/2}$. Let $f_i : D \to E$ be holomorphic mappings and let a_i, b_i be points of D such that $p = f_1(a_1)$, $f_1(b_1) = f_2(a_2)$, $\ldots$, $f_k(b_k) = q$. By homogeneity of D, we may assume that $a_i = 0$ for all i. By inserting extra terms in this chain if necessary, we may assume also that $b_i \in D_{r/2}$ for all i. We set $p_0 = p$, $p_1 = f_1(b_1)$, $\ldots$, $p_k = f_k(b_k) = q$. We have two cases to consider. Consider first the case where at least one of the p_i's is not contained in $\pi^{-1}(B_s)$. Then it is easy to see

$$\sum_{i=1}^{k} d_D(0, b_i) \geq \sum_{i=1}^{k} d_E(f_i(0), f_i(b_i)) = \sum_{i=1}^{k} d_E(p_{i-1}, p_i)$$

$$\geq \sum_{i=1}^{k} d_M(\pi(p_{i-1}), \pi(p_i)) \geq s.$$

Consider next the case where all p_i's are in $\pi^{-1}(B_s)$. Then

$$\sum_{i=1}^{k} d_D(0, b_i) \geqq c \sum_{i=1}^{k} d_{D_r}(0, b_i)$$

$$\geqq c \sum_{i=1}^{k} d_F(\varphi(p_{i-1}), \varphi(p_i)) \geqq c\, d_F(p, q),$$

where $\varphi : U \times F \to F$ is the projection. This shows that

$$d_E(p, q) \geqq \min[s, c\, d_F(p, q)] > 0.$$

Thus E is hyperbolic.

Assume that M and F are complete hyperbolic. Let $\{p_n\}$ be a Cauchy sequence in E. Then $\{\pi(p_n)\}$ is a Cauchy sequence in M and therefore $\pi(p_n) \to x_0$ for some $x_0 \in M$. Choose a neighborhood U of x_0 in M such that $\pi^{-1}(U) = U \times F$. Choose $s > 0$, $r > 0$, and $c > 0$ as above. Given $\varepsilon > 0$ with $2\varepsilon < s$, choose an integer N such that $p_n \in \pi^{-1}(B_\varepsilon)$ and $d_E(p_n, p_m) < \varepsilon$ for $n, m > N$. We shall show that $d_F(\varphi(p_n), \varphi(p_m)) < \varepsilon/c$ for $n, m > N$. We fix $n, m > N$. We choose holomorphic mappings $f_i : D \to E$ and points $b_i \in D$ such that

$$p_n = f_1(0), \quad f_1(b_1) = f_2(0), \quad \ldots, \quad f_{k-1}(b_{k-1}) = f_k(0), \quad f_k(b_k) = p_m,$$

and $\sum_{i=1}^{k} d_D(0, b_i) < \varepsilon$. We may again assume without loss of generality that $b_i \in D_{r/2}$ for all i. Since

$$\varepsilon > \sum_{i=1}^{k} d_D(0, b_i) \geqq \sum_{i=1}^{k} d_E(f_i(0), f_i(b)) \geqq \sum_{i=1}^{k} d_M(\pi \circ f_i(0), \pi \circ f_i(b)),$$

it follows that $f_i(0) \in \pi^{-1}(B_{2\varepsilon}) \subset \pi^{-1}(B_s)$. Hence,

$$\varepsilon > \sum_{i=1}^{k} d_D(0, b_i) \geqq c \sum_{i=1}^{k} d_{D_r}(0, b_i)$$

$$\geqq c \sum_{i=1}^{k} d_F(\varphi \circ f_i(0), \varphi \circ f_i(b_i))$$

$$\geqq c\, d_F(\varphi(p_n), \varphi(p_m)).$$

This shows that $\{\varphi(p_n)\}$ is a Cauchy sequence in F, and therefore $\varphi(p_n) \to y_0$ for some $y_0 \in F$. Clearly, $p_n \to (x_0, y_0) \in U \times F$. Thus E is complete.   QED.

Remark. Since F may be considered as a closed complex submanifold of E, it follows that if E is (complete) hyperbolic, then F is also (complete) hyperbolic. On the other hand, M need not be hyperbolic even if both E and F are complete hyperbolic. Let $B^* = \{(z, w) \in \mathbf{C}^2;$ $0 < |z|^2 + |w|^2 < 1\}$. Then B^* is a holomorphic fiber bundle over $P_1(\mathbf{C})$ with fiber $D^* = \{z \in \mathbf{C}; 0 < |z| < 1\}$. The bundle space B^* is hyperbolic and the fiber D^* is complete hyperbolic, but the base space $P_1(\mathbf{C})$ is not hyperbolic. If we set $E = \{(z, w) \in \mathbf{C}^2; z \neq 0$ and $|z|^2 + |w|^2 < 1\}$, then E is a fiber bundle over $\mathbf{C}$ with fiber D^*. This furnishes an example where E is even complete.

By a reasoning similar to the proof of Theorem 4.14, Kiernan [4] proves that if E is a Hermitian vector bundle over a hyperbolic manifold M, then the open unit-ball bundle $\{X \in E; \| X \| < 1\}$ is hyperbolic.

We conclude this section with another example of a complete hyperbolic manifold.

THEOREM 4.15.   *A Siegel domain of the second kind is complete hyperbolic.*

Proof. In § 5 of Chapter II we proved that a Siegel domain of the second kind is equivalent to a domain contained in a product of balls. From that proof it is not difficult to see that a Siegel domain of the second kind can be written as the intersection of (possibly uncountably many) domains, each of which is biholomorphic to a product of balls. But a product of balls is complete hyperbolic. Our assertion follows now from Proposition 4.3.   QED.

5. *On Completeness of an Invariant Distance*

We saw in Theorem 4.7 a similarity between the pseudodistance d_M and a Riemannian metric on M. We shall point out here another important similarity.

Given any subset A of a complex manifold M and a positive number r, let $U(A; r)$ be the open set defined by

$$U(A; r) = \{p \in M; d_M(p, a) < r \text{ for some point } a \in A\}.$$

With this notation, we have

PROPOSITION 5.1. *Let o be a point of a complex manifold M and let r and r′ be positive numbers. Then*

$$U[U(o; r); r'] = U(o; r + r').$$

Proof. The inclusion $U[U(o; r); r'] \subset U(o; r + r')$ is true for any pseudodistance and makes use of the triangular axiom only. In order to prove the inclusion in the opposite direction, let $p \in U(o; r + r')$ and set $d_M(o, p) = r + r' - 3\varepsilon$. Then there are points $a_i, b_i \in D$ and holomorphic mappings $f_i : D \to M$, $i = 1, \ldots, k$, such that

$$f_1(a_1) = o,$$
$$f_i(b_i) = f_{i+1}(a_{i+1}) \quad \text{for} \quad i = 1, \ldots, k-1,$$
$$f_k(b_k) = p,$$
$$\sum_{i=1}^{k} \varrho(a_i, b_i) < r + r' - 2\varepsilon.$$

Let j be the largest integer, $1 \le j \le k$, such that

$$\sum_{i=1}^{j-1} \varrho(a_i, b_i) < r - \varepsilon.$$

Let c_j be the point on the geodesic from a_j to b_j in D such that

$$\sum_{i=1}^{j-1} \varrho(a_i, b_i) + \varrho(a_j, c_j) = r - \varepsilon.$$

If we set

$$q = f_j(c_j),$$

then $d_M(o, q) < r$ and $d_M(q, p) < r'$ so that $p \in U(q; r') \subset U[U(o; r); r']$.   QED.

In the proof of Theorem 4.7 we promised to show that if a hyperbolic manifold M is complete in the sense that every Cauchy sequence converges, then the closure $\bar{U}(o; r)$ of $U(o; r)$ is compact for all $o \in M$ and all positive numbers r. We shall show that this assertion is true for a larger class of metric spaces.

THEOREM 5.2. *Let M be a locally compact metric space with distance function d satisfying the equality*

$$U[U(o; r); r'] = U(o; r + r')$$

for all $o \in M$ and all positive numbers r and r'. Then M is complete in the sense that every Cauchy sequence converges if and only if the closure $\bar{U}(o; r)$ of $U(o; r)$ is compact for all $o \in M$ and all positive numbers r.

Proof. We have only to prove that if M is (Cauchy-) complete, then $\bar{U}(o; r)$ is compact; the implication in the opposite direction is trivial.

LEMMA. *$\bar{U}(o; r)$ is compact if there exists a positive number b such that $\bar{U}(p; b)$ is compact for every $p \in U(o; r)$.*

Proof of Lemma. Since M is locally compact, there is a positive number $s < r$ such that $\bar{U}(o; s)$ is compact. It suffices to show that if $\bar{U}(o; s)$ is compact, so is $\bar{U}[o; s + (b/2)]$. Let $p_1, p_2, \ldots$ be points of $\bar{U}[o; s + (b/2)]$. Let $q_1, q_2, \ldots$ be points of $\bar{U}(o; s)$ such that $d(p_i, q_i) < \frac{3}{4}b$. Since $\bar{U}(o; s)$ is compact, we may assume (by choosing a subsequence if necessary) that $q_1, q_2, \ldots$ converges to some point, say q, of $\bar{U}(o; s)$. Then $\bar{U}(q; b)$ contains all p_i for large i. Since $\bar{U}(q; b)$ is compact by assumption, a suitable subsequence of $p_1, p_2, \ldots$ converges to a point p of $\bar{U}(q; b)$. Since $p_1, p_2, \ldots$ is a sequence in $\bar{U}[o; s + (b/2)]$ and $\bar{U}[o; s + (b/2)]$ is a closed set, the limit point p lies in $\bar{U}[o; s + (b/2)]$. This completes the proof of Lemma.

The proof of Theorem 5.2 is now reduced to showing that there exists a positive number b such that $\bar{U}(p; b)$ is compact for all $p \in M$. Assume the contrary. Then there exists a point $p_1 \in M$ such that $\bar{U}(p_1; \frac{1}{2})$ is noncompact. Applying Lemma to $\bar{U}(p_1; \frac{1}{2})$ we see that there exists a point $p_2 \in \bar{U}(p_1; \frac{1}{2})$ such that $\bar{U}[p_2; (1/2^2)]$ is noncompact. In this way we obtain a Cauchy sequence $p_1, p_2, p_3, \ldots$ such that $p_k \in \bar{U}[p_{k-1}; (1/2^{k-1})]$ and $\bar{U}[p_k; (1/2^k)]$ is noncompact. Let p be the limit point of the Cauchy sequence $p_1, p_2, \ldots$. Since M is locally compact, for a suitable positive number c, $\bar{U}(p; c)$ is compact. For a sufficiently large k, $\bar{U}[p_k; (1/2^k)]$ is a closed set contained in $\bar{U}(p; c)$ and hence must be compact. This is a contradiction.   QED.

Remark. Let M be a hyperbolic manifold and M^* its completion with respect to the distance d_M. Unfortunately, M^* need not be locally compact and hence it need not be complete in the strong sense that every closed ball of radius r in M^* is compact. An example of a hyperbolic manifold M such that M^* is not locally compact has been found by Kiernan [4]. A similar example of a Riemannian manifold has been found by D. B. A. Epstein (see Kobayashi [4]).

V

Holomorphic Mappings into Hyperbolic Manifolds

1. The Little Picard Theorem

The classical theorem of Liouville states that every bounded holomorphic function on the entire complex plane $\mathbf{C}$ is a constant function. We may prove this using the distance-decreasing property of the Carathéodory distance. Let D be a bounded domain and $f : \mathbf{C} \to D$ a holomorphic mapping. Since the Carathéodory pseudodistance $c_{\mathbf{C}}$ for $\mathbf{C}$ is trivial and the Carathéodory pseudodistance c_D for D is a distance, the distance-decreasing property of f implies that f is a constant mapping. The same reasoning applies to the invariant pseudodistance d_M. Delete two points a and b from $\mathbf{C}$. We know (see Corollary 4.12 of Chapter IV) that $\mathbf{C} - \{a, b\}$ is hyperbolic. On the other hand, $d_{\mathbf{C}}$ is trivial. Hence, every holomorphic mapping $f : \mathbf{C} \to \mathbf{C} - \{a, b\}$ is a constant mapping; in other words, every entire function with two lacunary values must be a constant function. This is the so-called little Picard theorem. We may now state

THEOREM 1.1. *Let M' be a complex manifold on which a complex Lie group acts transitively. Let M be a hyperbolic manifold. Then every holomorphic mapping $f : M' \to M$ is a constant mapping.*

The fact that the pseudodistance $d_{M'}$ for such a manifold M' is trivial was established in § 1 of Chapter IV (see Example 3).

2. The Automorphism Group of a Hyperbolic Manifold

The following lemma is due to van Dantzig and van der Waerden [1]. For the proof, see also Kobayashi and Nomizu [1, pp. 46–50].

LEMMA. *The group $I(M)$ of isometries of a connected, locally compact metric space M is locally compact with respect to the compact-open topology, and its isotropy subgroup $I_p(M)$ is compact for each $p \in M$. If M is moreover compact, then $I(M)$ is compact.*

From this lemma, we derive the following

THEOREM 2.1. *The group $H(M)$ of holomorphic transformations of a hyperbolic manifold M is a Lie transformation group, and its isotropy subgroup $H_p(M)$ at $p \in M$ is compact. If M is moreover compact, then $H(M)$ is finite.*

Proof. Let $I(M)$ the group of isometries of M with respect to the invariant distance d_M. Since $H(M)$ is a closed subgroup of $I(M)$, it follows from Lemma that $H(M)$ is locally compact with respect to the compact-open topology and $H_p(M)$ is compact for $p \in M$. By a theorem of Bochner and Montgomery [1], a locally compact group of differentiable transformations of a manifold is a Lie transformation group. Hence, $H(M)$ is a Lie transformation group, thus proving the first assertion. Assume that M is compact. Another theorem of Bochner and Montgomery [2] states that the group of holomorphic transformations of a compact complex manifold is a complex Lie transformation group. The second assertion follows from the following theorem.

THEOREM 2.2. *A connected complex Lie group of holomorphic transformations acting effectively on a hyperbolic manifold M reduces to the identity element only.*

Proof. Assume the contrary. Then a complex one-parameter subgroup acts effectively and holomorphically on M. Its universal covering group $\mathbf{C}$ acts essentially effectively on M. For each point $p \in M$, this action defines a holomorphic mapping $z \in \mathbf{C} \to z(p) \in M$. By Theorem 1.1, this holomorphic mapping $\mathbf{C} \to M$ must be a constant mapping. Since the identity element $0 \in \mathbf{C}$ maps p into p, every element z of $\mathbf{C}$ maps p into p. Since p is an arbitrary point of M, this shows that the action of $\mathbf{C}$ on M is trivial, contradicting the assumption.   QED.

COROLLARY 2.3. *Let M be a Hermitian manifold whose holomorphic sectional curvature is bounded above by a negative constant. Then the group $H(M)$ of holomorphic transformations of M is a Lie transformation group, and its isotropy subgroup $H_p(M)$ is compact for every $p \in M$. No complex Lie transformation group of positive dimension acts nontrivially on M. If M is moreover compact, then $H(M)$ is a finite group.*

Proof. This is immediate from Theorem 4.11 of Chapter IV and from Theorems 2.1 and 2.2.   QED.

We shall explain Theorem 2.2 from a slightly different viewpoint. Let X be a holomorphic vector field on a hyperbolic manifold M. Although it generates a local one-parameter group of local holomorphic transformations of M, it may in general not generate a global one-parameter group of holomorphic transformations. If it does, we call X a complete vector field. If X is holomorphic, so is JX, where J denotes the complex structure of M. Theorem 2.2 means that if X is complete, then JX can not be complete.

For a bounded domain in $\mathbf{C}^n$, Theorems 2.1 and 2.2 have been proved by H. Cartan [1]. (Of course, the second assertion of Theorem 2.1 is meaningless for a bounded domain.) We note that for a bounded domain these results of H. Cartan may be obtained by means of its Carathéodory distance. For a bounded domain, the Bergman metric may be also used to prove these results. Unlike the Carathéodory distance, the Bergman metric may be constructed even on some compact complex manifolds, e.g., any algebraic hypersurface of degree $d > n + 2$ in $P_{n+1}(\mathbf{C})$. For more details, see Kobayashi [1]. It would be of some interest to note here that one does not know if a generic algebraic hypersurface of degree $d > n + 2$ in $P_{n+1}(\mathbf{C})$ is hyperbolic or not (cf. Problem 4 in § 3 of Chapter IX). Theorem 2.2 is essentially equivalent to Theorem D (or Theorem D') in Wu [2]. For a systematic account of the holomorphic transformation group of a complex manifold, see Kaup [1].

From Corollary 4.13 of Chapter IV and Theorem 2.2 we obtain the following result of Schwarz and Klein (see Schwarz [1] and Poincaré [1]).

COROLLARY 2.4. *If M is a compact Riemann surface of genus $g \geqq 2$, then the group $H(M)$ of holomorphic transformations of M is finite.*

Hurwitz proved that the order of $H(M)$ does not exceed $84(g - 1)$. For similar results on algebraic surfaces, see Andreotti [1]. The following result is originally due to Bochner [1], Hawley [1], and Sampson [1].

COROLLARY 2.5. *Let M be a compact complex manifold that has a bounded domain $D \subset \mathbf{C}^n$ as a covering manifold so that $M = D/\Gamma$, where Γ is a properly discontinuous group of holomorphic transformations acting freely on D. Then the group H(M) of holomorphic transformations of M is finite.*

Proof. Since D is hyperbolic (Corollary 4.5 of Chapter IV), M is also hyperbolic by Theorem 4.7 of Chapter IV. Our assertion now follows from Theorem 2.2.   QED.

It is possible to obtain Corollaries 2.4 and 2.5 from the following result (Kobayashi [5]): *If M is a compact complex manifold whose first Chern class $c_1(M)$ is negative, then the group H(M) of holomorphic transformations of M is finite.* A compact Kaehler manifold M with negative Ricci tensor satisfies $c_1(M) < 0$.

THEOREM 2.6. *Let M be a hyperbolic manifold of complex dimension n. Then the group H(M) of holomorphic transformations of M has dimension $\leq 2n + n^2$ and the linear isotropy representation of the isotropy subgroup $H_p(M)$, $p \in M$, is faithful and is contained in $U(n)$.*

Proof. Let $\varrho : H_p(M) \to GL(n; \mathbf{C})$ be the linear isotropy representation. Since $H_p(M)$ is compact by Theorem 2.2, its image $\varrho[H_p(M)]$ is compact and is contained in $U(n)$ [after a suitable change of basis in the tangent space $T_p(M)$]. Since $H_p(M)$ is compact, we can find a Riemannian metric on M invariant by $H_p(M)$ so that $H_p(M)$ is contained in the group of isometries of M with respect to that Riemannian metric. It follows then that ϱ is faithful. Hence, dim $H_p(M) \leq$ dim $U(n) = n^2$. Finally, dim $H(M) \leq$ dim $M +$ dim $H_p(M) \leq 2n + n^2$.   QED.

Remark. It is not difficult to see that if dim $H(M)$ attains its maximum $2n + n^2$ in Theorem 2.6, then M is biholomorphic to the open unit ball in $\mathbf{C}^n$. We shall only outline its proof. The assumption dim $H(M) = 2n + n^2$ implies that $H_p(M) = U(n)$ for each $p \in M$ and $H(M)$ is transitive on M. Then there is a Hermitian metric on M which is invariant by $H(M)$. Since the isotropy subgroup $H_p(M) = U(n)$ contains the element $-I$, there is no nonzero tensor of odd degree (i.e., odd number of indices) at p, which is invariant by $H_p(M)$. In particular, the torsion tensor field of the invariant Hermitian metric must vanish. Hence, the metric is Kaehlerian. Since $U(n)$ is transitive on the unit sphere, the holomorphic sectional curvature of this invariant Kaehler metric is a constant. Since M is hyperbolic, this constant must be negative. Since a homogeneous Riemannian manifold of negative curvature is simply connected (see Kobayashi and Nomizu [1, Vol. II, p. 105]), M is simply connected. Being a simply connected complete Kaehler manifold of negative constant holomorphic sectional curvature, M is biholomorphic with the open unit ball in $\mathbf{C}^n$, (see Kobayashi and Nomizu [1, Vol. II, pp. 169–170]).

3. *Holomorphic Mappings into Hyperbolic Manifolds*

The following theorem is similar to the result of Van Dantzig and Van der Waerden (see § 2), but it is easier to prove.

THEOREM 3.1.   *Let M be a connected, locally compact, separable space with pseudodistance d_M and N a connected, locally compact, complete metric space with distance d_N. The set F of distance-decreasing mappings $f : M \to N$ is locally compact with respect to the compact-open topology. In fact, if p is a point of M and K is a compact subset of N, then the subset $F(p, K) = \{f \in F; \ f(p) \in K\}$ of F is compact.*

Proof.   Let $\{f_n\}$ be a sequence of mappings belonging to $F(p, K)$. We shall show that a suitable subsequence converges to an element of $F(p, K)$. We take a countable set $\{p_i\}$ of points which is dense in M. We set $K_i = \{q \in N; \ d_N(q, K) \leq d_M(p, p_i)\}$. Then K_i is a closed $d_M(p, p_i)$-neighborhood of K and hence is compact. Since each f_n is distance-decreasing so that $f_n(p_i)$ is in the compact set K_i for every n. By the standard argument of taking the diagonal subsequence, we can choose a subsequence $\{f_{n_k}\}$ such that $\{f_{n_k}(p_i)\}$ converges to some point of K_i for each p_i as k tends to infinity. By changing the notation, we denote this subsequence by $\{f_n\}$ so that $\{f_n(p_i)\}$ converges for each p_i. We now want to show that $\{f_n(q)\}$ converges for each $q \in M$. We have

$$d_N(f_n(q), f_m(q)) \leq d_N(f_n(q), f_n(p_i)) + d_N(f_n(p_i), f_m(p_i))$$
$$+ d_N(f_m(p_i), f_m(q))$$
$$\leq 2d_M(q, p_i) + d_N(f_n(p_i), f_m(p_i)).$$

Given any positive number ε, we choose p_i such that $2d_M(q, p_i) < \varepsilon/2$. We choose also an integer n_0 such that $d_N(f_n(p_i), f_m(p_i)) < \varepsilon/2$ for $n, m > n_0$. Then $d_N(f_n(q), f_m(q)) < \varepsilon$ for $n, m > n_0$. This shows that

$\{f_n(q)\}$ is a Cauchy sequence for each q. Since N is complete, we may define a mapping $f : M \to N$ by

$$f(q) = \lim_{n \to \infty} f_n(q).$$

Since each f_n is distance-decreasing, so is f. Since $f_n(p) \in K$ for each n, it follows that f maps p into K. We shall complete the proof by showing that the convergence $f_n(q) \to f(q)$ is uniform on every compact subset C of M. Let $\delta > 0$ be given. For each $q \in C$, choose an integer n_q such that $d_N(f_n(q), f(q)) < \delta/4$ for $n > n_q$. Let U_q be the open $\delta/4$-neighborhood of q in M. Then for any $x \in U_q$ and $n > n_q$, we have

$$d_N(f_n(x), f(x)) \leq d_N(f_n(x), f_n(q)) + d_N(f_n(q), f(q)) + d_N(f(q), f(x))$$

$$\leq 2d_M(x, q) + \frac{\delta}{4} < \delta.$$

Now C can be covered by a finite number of U_q's, say $U_i = U_{q_i}$, $i = 1$, $\ldots, s$. It follows that if $n > \max_i\{n_{q_i}\}$, then

$$d_N[f_n(x), f(x)] < \delta \qquad \text{for each} \quad x \in C.$$

QED.

Comparing this proof with that of Lemma for Theorem 2.1, we find that Theorem 3.1 is easier for the following reasons: In Theorem 3.1, N is assumed to be complete and there is no need to prove the existence of f^{-1}.

As an application of Theorem 3.1 we have

THEOREM 3.2.   *Let M be a complex manifold and N a complete hyperbolic manifold. Then the set F of holomorphic mappings $f : M \to N$ is locally compact with respect to the compact-open topology. For a point p of M and a compact subset K of N, the subset $F(p, K) = \{f \in F; f(p) \in K\}$ of F is compact.*

Proof.   This follows from Theorem 3.1 and from the fact that if a sequence of holomorphic mappings $f_n \in F$ converges to a continuous mapping f in the compact-open topology, then f is holomorphic.   QED.

THEOREM 3.3.   *Let M be a hyperbolic manifold and o a point of M. Let $f : M \to M$ be a holomorphic mapping such that $f(o) = o$. We denote by $df_o : T_o(M) \to T_o(M)$ the differential of f at o. Then*

(1) *The eigenvalues of df_o have absolute value ≤ 1;*

(2) *If df_o is the identity linear transformation, then f is the identity transformation of M;*

(3) *If $|\det f_o| = 1$, then f is a biholomorphic mapping.*

Proof. Let r be a positive number such that the closed r-neighborhood $B_r = \{p \in M; \ d_M(o, p) \leq r\}$ of o is compact. Let F_o denote the set of all (holomorphic or not) mappings of B_r into itself which leave o fixed and are distance-decreasing with respect to d_M. By Theorem 3.1, F_o is compact.

Let $f : M \to M$ be a holomorphic mapping such that $f(o) = o$. Let λ be an eigenvalue of df_o. For each positive integer k, the mapping f^k, restricted to B_r, belongs to F_o and its differential $(df_o)^k$ at o has eigenvalue λ^k. If $|\lambda| > 1$, then $|\lambda^k|$ goes to infinity as k goes to infinity, in contradiction to the fact that F_o is compact. This proves (1).

For the sake of simplicity, we denote by $d^m f_o$ all partial derivatives of order m at o. We want to show that if df_o is the identity transformation of $T_o(M)$, then $d^m f_o = 0$ for $m \geq 2$. Let m be the least integer ≥ 2 such that $d^m f_o \neq 0$. Then $d^m(f^k)_o = k d^m f_o$ for all positive integers k. As k goes to infinity, $d^m(f^k)_o$ also goes to infinity, in contradiction to the fact that F_o is compact. This proves (2).

Assume $|\det df_o| = 1$. From (1), it follows that all eigenvalues of df_o have absolute value 1. Put df_o in Jordan's canonical form. We claim that df_o is then in diagonal form. If it is not, it must have a diagonal block of this form:

$$\begin{bmatrix} \lambda & 1 & & 0 \\ & \lambda & \cdot & \\ & & \cdot & \cdot \\ & & & \cdot & 1 \\ 0 & & & & \lambda \end{bmatrix}, \qquad |\lambda| = 1.$$

The corresponding diagonal block of $(df_o)^k$ is then of the form

$$\begin{bmatrix} \lambda^k & k\lambda^{k-1} & & * \\ & \cdot & \cdot & \\ & & \cdot & \cdot & k\lambda^{k-1} \\ & & & \cdot & \\ 0 & & & & \lambda^k \end{bmatrix}.$$

It follows that the entries $k\lambda^{k-1}$ immediately above the diagonal of $(df_o)^k$ diverge to infinity as k goes to infinity, contradicting the compactness of F_o.

Since df_o is a diagonal matrix whose diagonal entries have absolute value 1, there is a subsequence $\{(df_o)^{k_i}\}$ of $\{(df_o)^k\}$ such that $(df_o)^{k_i}$ converges to the identity matrix as i goes to infinity. We denote now by F_o the set of all holomorphic mappings from M into itself leaving the point o fixed. Assume that M is complete. By Theorem 3.2, F_o is compact. Then there is a subsequence of $\{f^{k_i}\}$ which converges to a holomorphic mapping $h \in F_o$. By change of notations, we may assume that $\{f^{k_i}\}$ is this subsequence. Then $dh_o = \lim d(f^{k_i})_o$ is the identity transformation of $T_o(M)$. By (2) of the present theorem, h must be the identity transformation of M. Since F_o is compact, there is a convergent subsequence of $\{f^{k_i-1}\}$. By change of notations, we may assume that $\{f^{k_i-1}\}$ converges to a holomorphic mapping $g \in F_o$ as i goes to infinity. Then

$$f \circ g = f \circ (\lim f^{k_i-1}) = \lim f^{k_i} = \text{identity transformation.}$$

Similarly, $g \circ f$ is the identity transformation of M. This shows that f is a biholomorphic mapping with inverse g.

If M is not complete, we argue as follows. Let r be a small positive number such that the closed r-neighborhood B_r of o is compact. We denote again by F_o the set of all mappings of B_r into itself which leave the point o fixed and are distance-decreasing with respect to d_M. By Theorem 3.1, F_o is compact. We obtain a sequence $\{f^{k_i}\}$ which converges to an element h of F_o. Since the convergence is uniform, h is holomorphic in the interior of B_r and $dh_o = \lim d(f^{k_i})_o$ is the identity transformation of $T_o(M)$. From the proof of (2) above, we see that h is the identity transformation of B_r. Let W be the largest open subset of M with the property that some subsequence of $\{f^{k_i}\}$ converges to the identity transformation on W. (To prove the existence of W, consider the union $W = \cup W_j$ of all open subsets W_j of M such that on each W_j some subsequence of $\{f^{k_i}\}$ converges to the identity transformation. A countable number of W_j's already cover W. We consider the corresponding countable number of subsequences of $\{f^{k_i}\}$ and can extract a desired subsequence of f^{k_i} for W by the standard argument using the diagonal subsequence.) Without loss of generality, we may assume that f^{k_i} converges to the identity transformation on W. Since W contains the interior of B_r, it is nonempty. Let $p \in \bar{W}$ and U a neighborhood of p with compact closure $\bar{U}$. Since $\lim f^{k_i}(p) = p$ and each f^{k_i} is distance-decreasing, there is a neighborhood V of p such that $f^{k_i}(V) \subset U$ for $i \geq i_0$. Let F be the set of all distance-decreasing mappings from V into $\bar{U}$. By Theorem 3.1, F is compact. We extract a subsequence from $\{f^{k_i}\}$ which is convergent on V. Since it has to converge to the identity transformation on $V \cap W$, it converges to the identity transformation on V. This proves that W is closed and hence $W = M$. The remainder of the proof is the same as the case where M is complete.   QED.

For a bounded domain, Theorem 3.3 is due to H. Cartan [2, 3] (and also to Carathéodory [3]). Theorem 3.3 has been proved by Kaup [1] under weaker conditions. It has been also proved independently by Wu [2] under slightly stronger conditions.

Let M be a domain in $\mathbf{C}^n$ and $\bar{D}$ the closed unit disk in $\mathbf{C}$. According to Oka [1], M is said to be *pseudoconvex* if every continuous mapping $f : \bar{D} \times [0, 1] \rightarrow \mathbf{C}^n$ such that

(1) for each $t \in [0, 1]$ the mapping $f_t : D \rightarrow \mathbf{C}^n$ defined by $f_t(z) = f(z, t)$ is holomorphic and

(2) $f(z, t) \in M$ unless $|z| < 1$ and $t = 1$,

maps $\bar{D} \times [0, 1]$ necessarily into M.

For other definitions of pseudoconvexity and their equivalence with this definition, see Bremermann [2] and Lelong [1]. As an application of Theorem 3.2 we shall prove the following theorem.

THEOREM 3.4. *If a domain M in $\mathbf{C}^n$ is complete hyperbolic, then it is pseudoconvex.*

Proof. Let ∂D denote the boundary of $\bar{D}$, i.e., $\partial D = \bar{D} - D$. Since $f(\partial D \times [0, 1])$ is a compact subset of M, there exists a compact neighborhood K of $f(\partial D \times [0, 1])$ in M. Then we can find a point $z_0 \in D$ such that $f(z_0, t) \in K$ for $0 \leq t \leq 1$. By Theorem 3.2, the family $F(z_0, K)$ of holomorphic mappings $D \rightarrow M$ which send z_0 into K is compact. Since each f_t is in $F(z_0, K)$ for $0 \leq t < 1$, the limit mapping f_1 must be also in $F(z_0, K)$. In particular, f_1 maps D into M, i.e., $f(z, 1) \in M$ for $z \in D$.   QED.

Let B_n be the unit open ball in $\mathbf{C}^n$. From Theorem 3.4 we see that $B_n - \{0\}$ (where 0 denotes the origin) is not complete hyperbolic for $n \geq 2$.

VI

The Big Picard Theorem and
Extension of Holomorphic Mappings

1. Statement of the Problem

The classical big Picard theorem is usually stated as follows:

If a function $f(z)$ holomorphic in the punctured disk $0 < |z| < R$ has an essential singularity at $z = 0$, then there is at most one value a ($\neq \infty$) such that the equation $f(z) = a$ has only a finite number of solutions in the disk.

We may rephrase the statement above as follows. If $f(z) = a$ has only a finite number of solutions in the disk, then it has no solutions in a smaller disk $0 < |z| < R'$, $R' \leq R$. Hence, the big Picard theorem says that *if a function $f(z)$ holomorphic in the punctured disk $0 < |z| < R$ misses two values a, b ($\neq \infty$), then it has a removable singularity or a pole at $z = 0$, i.e., it can be extended to a meromorphic function on the disk $|z| < R$.*

In other words, *if f is a holomorphic mapping from the punctured disk $0 < |z| < R$ into $M = P_1(\mathbf{C}) - \{3 \text{ points}\}$, then f can be extended to a holomorphic mapping from the disk $|z| < R$ into $P_1(\mathbf{C})$.*

We consider the following problem in this section.

Let Y be a complex manifold and M a submanifold which is hyperbolic and relatively compact. Given a holomorphic mapping f from the punctured disk $0 < |z| < R$ into M, is it possible to extend it to a holomorphic mapping from the disk $|z| < R$ into Y?

79

We shall give an affirmative answer in some special cases. We shall consider also the case where Y and M are complex spaces and the domain is of higher dimension.

The following example by Kiernan [4] shows that the answer to the question above is in general negative. (It is slightly simpler than the original example of Kiernan.)

Let $Y = P_2(C)$ with homogeneous coordinate system (u, v, w). Let $M = \{(1, v, w) \in P_2(\mathbf{C}); 0 < |v| < 1, |w| < |e^{1/v}|\}$. Then the mapping $(1, v, w) \to (v, we^{1/v})$ defines a biholomorphic equivalence between M and $D^* \times D$, (where $D^* = \{z \in \mathbf{C}; 0 < |z| < 1\}$ and $D = \{z \in \mathbf{C}; |z| < 1\}$). Hence, M is hyperbolic. Let $f : D^* \to M$ be the mapping defined by

$$f(z) = (1, z, \tfrac{1}{2}e^{1/z}) \qquad z \in D^*.$$

Then f cannot be extended to a holomorphic mapping from D into $P_2(\mathbf{C})$.

2. *The Invariant Distance on the Punctured Disk*

Let D be the open unit disk in $\mathbf{C}$ and D^* the punctured disk $D - \{0\}$, i.e., $D^* = \{z \in \mathbf{C}; 0 < |z| < 1\}$. Let H be the upper half-plane $\{w = u + iv \in \mathbf{C}; v > 0\}$ in $\mathbf{C}$. The invariant metric $ds_H{}^2$ of curvature -1 on H is given by

$$ds_H{}^2 = \frac{dw\, d\bar{w}}{v^2}.$$

Let $p : H \to D^*$ be the covering projection defined by

$$z = p(w) = e^{2\pi i w} \qquad \text{for} \quad w \in H.$$

Let $ds_{D^*}^2$ be the metric on D^* defined by

$$p^*(ds_{D^*}^2) = ds_H{}^2.$$

Since $dz = 2\pi i z\, dw$ and $z\bar{z} = e^{-4\pi v}$, we obtain easily

$$ds_{D^*}^2 = \frac{4\, dz\, d\bar{z}}{z\bar{z}[\log(1/z\bar{z})]^2}.$$

For each positive number $r < 1$, let $L(r)$ denote the arc-length of the circle $|z| = r$ with respect to ds^2_{D*}. Then

$$L(r) = \frac{2\pi}{\log(1/r)}.$$

In the definition of the pseudodistance d_M in § 1 of Chapter IV, the distance ϱ on D is the one defined by the Poincaré–Bergman metric $ds_D{}^2$ of D. Without loss of generality (i.e., by multiplying a suitable positive constant to the metric), we may assume that $ds_D{}^2$ has curvature -1 so that H is not only biholomorphic but also isometric to D. Then the invariant distance d_H of H is the one defined by $ds_H{}^2$. By Proposition 1.6 of Chapter IV, the invariant distance d_{D*} coincides with the one defined by ds^2_{D*}. We state the result of this section in the form convenient for later uses.

PROPOSITION 2.1.   *Let $L(r)$ be the arc-length of the circle $|z| = r < 1$ with respect to the invariant distance d_{D*} of the punctured unit disk D^*. Then $L(r) = a/\log(1/r)$, where a is a constant. In particular,*

$$\lim_{r \to 0} L(r) = 0.$$

Since we want to consider not only complex manifolds but also complex spaces, we state the following proposition for a metric space M, not just for a hyperbolic manifold M.

PROPOSITION 2.2.   *Let M be a locally simply connected metric space with distance function d_M. Let D^* be the punctured unit disk with the invariant distance d_{D*}. Let $f : D^* \to M$ be a distance-decreasing mapping. Assume that there is a sequence of points $\{z_k\}$ in D^* such that $\lim_{k\to\infty} z_k = 0$ and $f(z_k)$ converges to a point $p_0 \in M$. Then for each positive $r < 1$, f maps the circle $|z| = r$ into a closed curve which is homotopic to zero.*

Proof.   Set $r_k = |z_k|$. Since the closed curves

$$\gamma_k(t) = f(r_k e^{2\pi i t}), \qquad 0 \le t \le 1, \quad k = 1, 2, \ldots$$

are all homotopic to each other, it suffices to prove that, for a sufficiently large integer k, the closed curve $f(\gamma_k)$ is homotopic to zero. Let U be a simply connected neighborhood of p_0 in M and take a smaller neighborhood V such that $\bar{V} \subset U$. Let N be an integer such that $f(z_n) \in V$

for $n > N$. Since f is distance-decreasing, Proposition 2.1 implies that the arc-length of the closed curve $f(\gamma_k)$ approaches zero as k goes to infinity. Hence, if $k > N$ is sufficiently large, $f(\gamma_k)$ is contained in U. Since U is simply connected, $f(\gamma_k)$ is homotopic to zero.   QED.

COROLLARY 2.3.   *Let M be a locally simply connected, compact metric space with distance function d_M. Let D^* be the punctured unit disk with the invariant distance d_{D^*}. If $f : D^* \to M$ is a distance-decreasing mapping, then f maps each circle $|z| = r < 1$ into a closed curve which is homotopic to zero.*

Let M, D^*, and f be as in Proposition 2.2 or Corollary 2.3. Let $\tilde{M}$ be a covering space of M with projection π. Then f can be lifted to a mapping $\tilde{f} : D^* \to \tilde{M}$ such that $f = \pi \circ \tilde{f}$.

THEOREM 2.4.   *Let M be a complex manifold which has a covering manifold $\tilde{M}$ with Carathéodory distance $c_{\tilde{M}}$. Let D^* be the punctured disk. Let $f : D^* \to M$ be a holomorphic mapping such that, for a suitable sequence of points $z_k \in D^*$ converging to the origin, $f(z_k)$ converges to a point $p_0 \in M$. Then f extends to a holomorphic mapping of the (complete) disk D into M.*

Proof.   Since $c_{\tilde{M}}$ is assumed to be a distance, $d_{\tilde{M}}$ is also a distance by Proposition 2.1 of Chapter IV and hence $\tilde{M}$ is hyperbolic. By Theorem 4.7 of Chapter IV, M is also hyperbolic. Since $f : D^* \to M$ is distance-decreasing with respect to d_{D^*} and d_M, we can lift f to a holomorphic mapping $\tilde{f} : D^* \to \tilde{M}$ as we have seen above. Then $\tilde{f}(z_k)$ converges to a point $\tilde{p}_0 \in \tilde{M}$ such that $\pi(\tilde{p}_0) = p_0$, where π is the projection $\tilde{M} \to M$. The mapping $f : D^* \to \tilde{M}$ is distance-decreasing with respect to the Carathéodory distances c_{D^*} and $c_{\tilde{M}}$. Since the disk D is the completion of D^* with respect to the Carathéodory distance c_{D^*} (see § 3 of Chapter IV), $\tilde{f}$ can be extended to a holomorphic mapping $\tilde{f} : D \to \tilde{M}$ such that $\tilde{f}(0) = \tilde{p}_0$. It follows that f can be extended to a holomorphic mapping $f : D \to M$ such that $f(0) = p_0$.   QED.

COROLLARY 2.5.   *Let M be a compact complex manifold which has a covering manifold $\tilde{M}$ with Carathéodory distance $c_{\tilde{M}}$. Then every holomorphic mapping from the punctured disk D^* into M can be extended to a holomorphic mapping from the disk D into M.*

Remark. Theorem 2.4 can be applied to a complex manifold M of the form $\tilde{M}/\Gamma$, where $\tilde{M}$ is a bounded domain in $\mathbf{C}^n$ and Γ is a properly discontinuous group of holomorphic transformations acting freely on $\tilde{M}$. If $M = \tilde{M}/\Gamma$ is compact, Corollary 2.5 applies.

From Theorem 2.4 we obtain also the following:

COROLLARY 2.6. *Let M be a complex submanifold of a complex manifold Y such that M is compact. Assume that M has a covering manifold $\tilde{M}$ with Carathéodory distance $c_{\tilde{M}}$. Then every holomorphic mapping f of the punctured disk D^* into M satisfies one of the following two conditions:*

(1) *f can be extended to a holomorphic mapping of the disk D into M;*
(2) *For every neighborhood N of the boundary $\partial M = M - \tilde{M}$ of M in $\tilde{M}$, there exists a neighborhood U of the origin in the disk D such that $f(U - \{0\}) \subset N$.*

If we set $Y = P_1(\mathbf{C})$ and $M = Y - \{3 \text{ points}\}$, then Corollary 2.6 yields the classical big Picard theorem immediately.

The results of this section are due essentially to Huber [2], who obtained a generalization of the big Picard theorem in the following form:

Let Y be a Riemann surface and M a domain of hyperbolic type. Then every holomorphic mapping $f : D^* \to M$ can be extended to a holomorphic mapping $f : D \to Y$.

This follows from Corollary 2.6 and from the fact that the boundary of M is contained in another subdomain of Y which is of hyperbolic type. But this last fact is known only for Riemann surfaces.

3. *Mappings from the Punctured Disk into a Hyperbolic Manifold*

In this section we shall prove the following theorem of Kwack [1], which generalizes Theorem 2.4.

THEOREM 3.1. *Let M be a hyperbolic manifold and D^* the punctured unit disk. Let $f : D^* \to M$ be a holomorphic mapping such that, for a suitable sequence of points $z_k \in D^*$ converging to the origin, $f(z_k)$ converges to a point $p_0 \in M$. Then f extends to a holomorphic mapping of the unit disk D into M.*

COROLLARY 3.2. *If M is a compact hyperbolic manifold, then every holomorphic mapping $f : D^* \to M$ extends to a holomorphic mapping of D into M.*

Proof.   As in § 2, we set

$$r_k = |z_k|,$$
$$\gamma_k(t) = f(r_k e^{2\pi i t}), \qquad 0 \leq t \leq 1, \quad k = 1, 2, \ldots.$$

In other words, γ_k is the image of the circle $|z| = r_k$ by f.

Let U be a neighborhood of p_0 in M with local coordinate system $w^1, \ldots, w^n$. We may assume that p_0 is at the origin of this coordinate system. Let ε be a small positive number and let V be the open neighborhood of p_0 defined by

$$V: \qquad |w^i| < \varepsilon \qquad i = 1, \ldots, n.$$

Taking ε sufficiently small, we may assume that $\bar{V} \subset U$. Let W be the neighborhood of p_0 defined by

$$W: \qquad |w^i| < \varepsilon/2 \qquad i = 1, \ldots, n.$$

The problem is to show that, for a suitable positive number δ, the small punctured disk $\{z \in D^*; |z| < \delta\}$ is mapped into U by f.

Since the diameter of γ_k approaches zero as k goes to infinity by Proposition 2.1, all but a finite number of γ_k's are contained in W. Without loss of generality we may assume that all γ_k's are in W. By taking a subsequence of $\{z_k\}$ if necessary, we may assume also that the sequence $\{r_k\}$ is monotone decreasing. Consider the set of integers k such that the image of the annulus $r_{k+1} < |z| < r_k$ by f is not entirely contained in W. If this set of integers is finite, then f maps a small punctured disk $0 < |z| < \delta$ into $\bar{W}$. Assuming that this set of integers is infinite, we shall obtain a contradiction.

By taking a subsequence, we may assume that, for every k, the image of the annulus $r_{k+1} < |z| < r_k$ by f is not entirely contained in W. For each k, let

$$R_k = \{z \in D^*; a_k < |z| < b_k\}$$

be the largest open annulus such that (1) $a_k < r_k < b_k$ and (2) f maps R_k into W. We set

$$\sigma_k(t) = a_k e^{2\pi i t}, \qquad 0 \leq t \leq 1,$$
$$\tau_k(t) = b_k e^{2\pi i t}, \qquad 0 \leq t \leq 1.$$

In other words, σ_k is the inner boundary of the annulus R_k and τ_k is the outer boundary of the annulus R_k. From the definition of a_k and b_k, it is clear that both $f(\sigma_k)$ and $f(\tau_k)$ are contained in $\overline{W}$ but not in W. By Proposition 2.1, the diameters of $f(\sigma_k)$ and $f(\tau_k)$ approach zero as k goes to infinity. By taking a subsequence if necessary, we may assume that the sequences $\{f(\sigma_k)\}$ and $\{f(\tau_k)\}$ converge to points q and q' of $\overline{W} - W$, respectively. Since p_0 is in W and both q and q' are on the boundary of W, the points q and q' are distinct from p_0. (The point q might coincide with the point q'.) By taking a new coordinate system around p_0 if necessary, we may assume that

$$w^1(q) \neq w^1(p_0) = 0, \qquad w^1(q') \neq w^1(p_0) = 0.$$

Let $f(z) = [f^1(z), \ldots, f^n(z)]$ be the local expression of f on $f^{-1}(U) \subset D$. Then

$$\lim_{k\to\infty} f^1(\sigma_k) = w^1(q),$$

$$\lim_{k\to\infty} f^1(\tau_k) = w^1(q'),$$

$$\lim_{k\to\infty} f^1(z_k) = w^1(p_0) = 0.$$

It follows that if k is sufficiently large, then

$$f^1(z_k) \notin f^1(\sigma_k) \cup f^1(\tau_k).$$

If k is sufficiently large, we can find a simply connected open neighborhood G_k of $w^1(q)$ in $\mathbf{C}$ such that $f^1(\sigma_k) \subset G_k$ and $f^1(z_k) \notin G_k$. We apply Cauchy's theorem to the holomorphic function $1/[w^1 - f^1(z_k)]$ and the closed curve $f^1(\sigma_k)$ in G_k. Then

$$\int_{f^1(\sigma_k)} \frac{dw^1}{w^1 - f^1(z_k)} = 0.$$

This may be rewritten as follows:

$$\int_{\sigma_k} \frac{f^{1\prime}(z)}{f^1(z) - f^1(z_k)}\, dz = 0,$$

where $f^{1\prime}(z) = df^1(z)/dz$. Similarly, if k is sufficiently large,

$$\int_{\tau_k} \frac{f^{1\prime}(z)}{f^1(z) - f^1(z_k)}\, dz = 0.$$

On the other hand, the principle of the argument applied to the function $f^1(z) - f^1(z_k)$ defined in a neighborhood of the annulus R_k which is bounded by the curve $\tau_k - \sigma_k$ yields the following equality:

$$\int_{\tau_k} \frac{f^{1\prime}(z)}{f^1(z) - f^1(z_k)}\, dz - \int_{\sigma_k} \frac{f^{1\prime}(z)}{f^1(z) - f^1(z_k)}\, dz = 2\pi i(N - P),$$

where N and P denote the numbers of zeros and poles of $f^1(z) - f^1(z_k)$ in R_k. In the present situation, $P = 0$ and $N \geq 1$. We have arrived finally at a contradiction.   QED.

From Theorem 3.1 we obtain also the following:

COROLLARY 3.3.   *Let M be a hyperbolic submanifold of a complex manifold Y such that $\bar{M}$ is compact. Then every holomorphic mapping f of the punctured disk D^* into M satisfies one of the following two conditions*:

(1) *f can be extended to a holomorphic mapping of the disk D into M*;
(2) *For every neighborhood of N of the boundary $\partial M = \bar{M} - M$ of M in $\bar{M}$, there exists a neighborhood U of the origin in the disk D such that $f(U - \{0\}) \subset N$.*

This generalizes Corollary 2.6.

4. *Holomorphic Mappings into Compact Hyperbolic Manifolds*

As an application of Theorem 3.1 we shall prove the following result of Kwack [1].

THEOREM 4.1.   *Let M be a compact hyperbolic manifold. Let X be a complex manifold of dimension m and A an analytic subset of X of dimension $\leq m - 1$. Then every holomorphic mapping f from $X - A$ into M can be extended to a holomorphic mapping from X into M.*

Proof.   We shall first show that it suffices to prove the theorem when A is a nonsingular complex submanifold of X. Let S be the set of singular points of A. Then S is an analytic subset of X and dim $S <$ dim A (see, for instance, Narasimhan [1, pp. 56–58]). Since $A - S$ is a nonsingular complex submanifold of $X - S$, we extend first f to a holomorphic mapping from $X - S$ into M. Since dim $S <$ dim A, we obtain the theorem by induction on the dimension of A.

We shall now assume that A is nonsingular. For each point a of A

we want to find a neighborhood U in X such that $f \mid U \cap (X - A)$ can be extended to a holomorphic mapping from U into M. We may therefore assume that X is a polydisk

$$D \times D^{m-1} = \{z, t^1, \ldots, t^{m-1}) \in \mathbf{C}^m; \mid z \mid < 1, \mid t^1 \mid < 1, \ldots, \mid t^{m-1} \mid < 1\},$$

and that A is contained in the subset defined by $z = 0$. For the sake of simplicity, we denote $(t^1, \ldots, t^{m-1})$ and $(z, t^1, \ldots, t^{m-1})$ by t and (z, t), respectively.

For each fixed $t \in D^{m-1}$, we have a holomorphic mapping f_t from the punctured disk D^* into M defined by $f_t(z) = f(z, t)$. Applying Corollary 3.2 to each f_t, we extend f_t to a holomorphic mapping $f_t : D \to M$ and we set $f(0, t) = f_t(0)$. We have to prove that this extended mapping $f : X \to M$ is holomorphic. By the Riemann extension theorem, it suffices to show that $f : X \to M$ is continuous at every point of A. To prove that f is continuous at $a \in A$, we may assume without loss of generality that a is the origin $(z, t) = (0, 0)$. We set $p = f(0, 0) \in M$. Let ε be any positive number. Let V be the ε-neighborhood of 0 in D with respect to the Poincaré distance $\varrho = d_D$. Let W be the ε-neighborhood of 0 in D^{m-1} with respect to the distance $d_{D^{m-1}}$. We shall show that f maps $V \times W$ into the 3ε-neighborhood of p in M with respect to the distance d_M. Let $(z, t) \in V \times W$. Since the restriction of f to $D \times \{0\}$ is holomorphic and since $f : D \times \{0\} \to M$ is distance-decreasing, we have

$$d_M(f(0, 0), f(z, 0)) \leqq d_D(0, z) < \varepsilon.$$

We shall first consider the case $z \neq 0$. Since $f : D^* \times D^{m-1} \to M$ is holomorphic and hence distance-decreasing, we have

$$d_M(f(z, 0), f(z, t)) \leqq d_{D^* \times D^{m-1}}((z, 0), (z, t))$$
$$= d_{D^{m-1}}(0, t) < \varepsilon,$$

where the equality is a consequence of Proposition 1.5 of Chapter IV. We have now

$$d_M(f(0, 0), f(z, t)) \leqq d_M(f(0, 0), f(z, 0)) + d_M(f(z, 0), f(z, t)) < 2\varepsilon,$$

provided that $z \neq 0$. We shall now consider a point $(0, t) \in V \times W$. Choose any $z \in V$ different from zero. Since $f : D \times \{t\} \to M$ is holomorphic and hence distance-decreasing, we have

$$d_M(f(z, t), f(0, t)) \leq d_D(z, 0) < \varepsilon.$$

Hence,

$$d_M(f(0, 0), f(0, t)) \leq d_M(f(0, 0), f(z, t)) + d_M(f(z, t), f(0, t))$$
$$< 2\varepsilon + \varepsilon = 3\varepsilon.$$

This proves that f is continuous at $(0, 0)$.   QED.

5. Holomorphic Mappings into Complete Hyperbolic Manifolds

In Theorem 4.1, if A is smaller, it suffices to assume that M is complete hyperbolic. Before we make an exact definition, we prove

PROPOSITION 5.1.   Let $D^m = \{(z^1, \ldots, z^m) \in \mathbf{C}^m;\ |z^j| < 1\ $ for $j = 1, \ldots, m\}$ and let A be a subset of $D^m = D \times D^{m-1}$ of the form $A = \{0\} \times A'$, where A' is nowhere dense in D^{m-1}. Then the distance $d_{D^m - A}$ is the restriction of the distance d_{D^m} to $D^m - A$.

Proof.   Let p and q be two points of $D^m - A$. Since the injection $D^m - A \to D^m$ is holomorphic and hence distance-decreasing, we have $d_{D^m - A}(p, q) \geq d_{D^m}(p, q)$. To prove the proposition, it suffices to show the opposite inequality for every pair of points (p, q) belonging to a dense subset of $(D^m - A) \times (D^m - A)$.

Let S be the subset of $(D^m - A) \times (D^m - A)$ consisting of pairs (p, q) for which there exist points $a, b \in D$ and a holomorphic mapping $f : D \to D^m - A$ such that $d_{D^m}(p, q) = d_D(a, b), f(a) = p$, and $f(b) = q$. If $(p, q) \in S$, then

$$d_{D^m - A}(p, q) = d_{D^m - A}(f(a), f(b)) \leq d_D(a, b) = d_{D^m}(p, q).$$

It suffices therefore to prove that S is a dense subset.

Let $p = (a^1, \ldots, a^m)$ and $q = (b^1, \ldots, b^m)$ be arbitrary points of $D^m - A$. To show that every neighborhood of (p, q) in $(D^m - A) \times (D^m - A)$ contains a point of S, we may assume without loss of generality that a^1, b^1 and 0 are mutually distinct, for the set of such pairs is dense in $(D^m - A) \times (D^m - A)$. The distance $d_{D^m}(p, q)$ is equal to the maximum of $d_D(a^j, b^j), j = 1, \ldots, m$, say $d_D(a^k, b^k)$ (see Example 1 in § 1 of Chapter IV). We set $a = a^k$ and $b = b^k$ so that $d_{D^m}(p, q)$

$= d_D(a, b)$. Since $d_D(a^j, b^j) \leq d_D(a, b)$ for $j = 1, \ldots, m$, there exist holomorphic mappings $f_j : D \to D$ such that $f_j(a) = a^j$ and $f_j(b) = b^j$ for $j = 1, \ldots, m$. Since $a^1 \neq b^1$, we may impose the additional condition that f_1 be injective. Then $f_1^{-1}(0)$ is either empty or a single point $c \in D$. If $f_1^{-1}(0)$ is empty, then the mapping $f : D \to D^m$ defined by $f(z) = (f_1(z), \ldots, f_m(z))$ sends D into $D^m - A$, since $f_1(z)$ never vanishes. In this case, (p, q) belongs to S. Assume $c = f_1^{-1}(0)$. Then the mapping $f : D \to D^m$ defined above maps D into $D^m - A$ if and only if $[f_2(c), \ldots, f_m(c)]$ is not in A'. If $f(D) \subset D^m - A$, then (p, q) belongs to S. So, we have only to consider the case $[f_2(c), \ldots, f_m(c)] \in A'$.

We assert that given a positive number ε there exists a positive number δ such that for any points $c^j \in D$ ($j = 2, \ldots, m$) with $d_D(c^j, f_j(c)) < \delta$ there exist automorphisms $h_j : D \to D$ satisfying

$$h_j(f_j(c)) = c^j, \quad d_D(a^j, h_j(a^j)) < \varepsilon, \quad \text{and} \quad d_D(b^j, h_j(b^j)) < \varepsilon.$$

We shall first complete the proof of the proposition and then come back to the proof of this assertion.

Given $\varepsilon > 0$, let $\delta > 0$ be as above. Since A' is nowhere dense in D^{m-1}, there exists a point $(c^2, \ldots, c^m) \in D^{m-1} - A'$ such that $d_D(c^j, f_j(c)) < \delta$ for $j = 2, \ldots, m$. Let $h_j : D \to D$ be as above. We consider the points $p' = [a^1, h_2(a^2), \ldots, h_m(a^m)]$ and $q' = [b^1, h_2(b^2), \ldots, h_m(b^m)]$ of D^m and the holomorphic mapping $f' : D \to D^m$ defined by $f'(z) = (f_1(z), h_2(f_2(z)), \ldots, h_m(f_m(z)))$. Since a^1, b^1, and 0 are mutually distinct, both p' and q' are in $D^m - A$. Since $h_j : D \to D$ is distance-preserving, we have $d_D(h_j(a^j), h_j(b^j)) = d_D(a^j, b^j)$ and hence

$$d_{D^m}(p', q') = \text{Max}\{d_D(a^j, b^j); j = 1, \ldots, m\} = d_{D^m}(p, q) = d_D(a, b).$$

Clearly, $f'(a) = p'$ and $f'(b) = q'$. This shows that (p', q') belongs to S. Since $d_D(a^j, h_j(a^j)) < \varepsilon$, we have $d_{D^m}(p, p') < \varepsilon$. Similarly, $d_{D^m}(q, q') < \varepsilon$. Thus, the ε-neighborhood of (p, q) in $(D^m - A) \times (D^m - A)$ with respect to $d_{D^m} \times d_{D^m}$ contains a point (p', q') of S. This completes the proof of the proposition except for the proof of the assertion made above.

To simplify the notations in the assertion above, we denote a^j, b^j, $f_j(c), c^j$, and h_j by a, b, c, c', and h. respectively. Then the assertion we have to prove reads as follows:

LEMMA. *Let* $a, b, c \in D$ *be given. Then for any* $\varepsilon > 0$ *there is a* $\delta > 0$ *such that for any* $c' \in D$ *with* $d_D(c, c') < \delta$ *there exists an automorphism* $h : D \to D$ *satisfying*

$$h(c) = c', \qquad d_D(a, h(a)) < \varepsilon, \qquad and \qquad d_D(b, h(b)) < \varepsilon.$$

In order to prove the lemma, it is more convenient to replace D by the upper half-plane H in $\mathbf{C}$. Given $\varepsilon > 0$, choose $\delta_1 > 0$ and $\delta_2 > 0$ such that

$$d_H(a, r(a + t)) < \varepsilon \qquad and \qquad d_H(b, r(b + t)) < \varepsilon$$

for any real numbers t and r such that $|t| < \delta_1$ and $|r - 1| < \delta_2$. The set $\{r(c + t); |t| < \delta_1, |r - 1| < \delta_2\}$ contains the δ-neighborhood of c for some $\delta > 0$. Given c' in the δ-neighborhood of c, we can find an automorphism $h : H \to H$ of the form $h(z) = r(z + t)$ such that $h(c) = c'$, $d_H(a, h(a)) < \varepsilon$ and $d_H(b, h(b)) < \varepsilon$.   QED.

As an application of Proposition 5.1 we prove

THEOREM 5.2.   *Let* M *be a complete hyperbolic manifold. Let* X *be a complex manifold of dimension* m, *and let* A *be a subset which is nowhere dense in an analytic subset, say* B, *of* X *with* $\dim B \leq m - 1$. *Then every holomorphic mapping* $f : X - A \to M$ *can be extended to a holomorphic mapping* $X \to M$.

Proof.   As in the proof of Theorem 4.1, we can reduce the proof to the special case where $X = D^m = \{(z^1, \ldots, z^m) \in \mathbf{C}^m; |z^j| < 1\}$ and B is the subset defined by $z^1 = 0$ so that A is of the form $A = \{0\} \times A'$, where A' is nowhere dense in D^{m-1}. Since $f : D^m - A \to M$ is distance-decreasing, f can be extended to a continuous mapping from the completion of the metric space $D^m - A$ into M. By Proposition 5.1, D^m is the completion of $D^m - A$ with respect to the distance $d_{D^m - A}$. By the Riemann extension theorem, the extended continuous mapping $f : D^m \to M$ is necessarily holomorphic.   QED.

Theorem 5.2 contains the following result of Kwack [1], which was proved by a different method.

COROLLARY 5.3.   *Let* M *be a complete hyperbolic manifold. Let* X *be a complex manifold of dimension* m, *and let* A *be an analytic subset of di-*

mension $\leqq m - 2$. *Then every holomorphic mapping* $f : X - A \to M$ *can be extended to a holomorphic mapping* $X \to M$.

For extension of a holomorphic mapping $X - A \to M$ where M has a covering space which is a Stein manifold, see Andreotti and Stoll [1].

6. *Holomorphic Mappings into Relatively Compact Hyperbolic Manifolds*

The following result is the best solution we can give to the problem stated in § 1 at the moment.

THEOREM 6.1.   *Let* Y *be a complex manifold and* M *a complex submanifold of* Y *satisfying the following three conditions*:

(1) M *is hyperbolic*;

(2) *the closure of* M *in* Y *is compact*;

(3) *given a point* p *of* $\bar{M} - M$ *and neighborhood* U *of* p *in* Y, *there exists a neighborhood* V *of* p *in* Y *such that* $\bar{V} \subset U$ *and the distance between* $M \cap (Y - U)$ *and* $M \cap V$ *with respect to* d_M *is positive.*

Then every holomorphic mapping f *from the punctured disk* D^* *into* M *can be extended to a holomorphic mapping from the disk* D *into* Y.

Proof.   Let $\{r_k\}$, $0 < r_k < 1$, be a monotone decreasing sequence with $\lim r_k = 0$. We consider $\{r_k\}$ as a sequence of points in D^* converging to the origin. Since $\bar{M}$ is compact, we may assume (by taking a subsequence if necessary) that $\{f(r_k)\}$ converges to a point p_0 of $\bar{M}$. The case where p_0 is in M has been already considered (see Theorem 3.1). We assume therefore that $p_0 \in \bar{M} - M$. Let U be a neighborhood of p_0 in Y. Let γ_k denote the circle $|z| = r_k$ in D^*. We claim that there exists an integer N such that

$$f(\gamma_k) \subset U \qquad \text{for} \qquad k > N.$$

To prove our claim, assume the contrary. By taking a subsequence if necessary, we may assume that each circle γ_k has a point z_k such that $f(z_k)$ is not in U. Taking again a subsequence if necessary, we may assume that the sequence $\{f(z_k)\}$ converges to a point q_0 of $\bar{M}$. In view of Theorem 3.1, we have only to consider the case where q_0 is in $\bar{M} - M$.

Since $f(z_k)$ is outside U, the limit point q_0 is also outside U. Taking U sufficiently small, we may assume that q_0 is outside $\bar{U}$ so that $Y - \bar{U}$ is an open neighborhood of q_0. By (3), there is a smaller neighborhood

V of q_0 such that $\overline{V} \cap \overline{U} = \emptyset$ and the distance between $M \cap U$ and $M \cap V$ with respect to d_M is a positive number δ. Then

$$d_M(f(r_k), f(z_k)) \geq \delta \qquad \text{for large } k,$$

because $f(r_k) \in U$ and $f(z_k) \in V$. On the other hand, the arc-length $L(\gamma_k)$ of γ_k tends to zero as k goes to infinity (see Proposition 2.1). We have

$$d_M(f(r_k), f(z_k)) \leq L[f(\gamma_k)] \leq L(\gamma_k).$$

This is a contradiction. We have thus shown that there exists an integer N such that

$$f(\gamma_k) \subset U \qquad \text{for } k > N.$$

The rest of the proof goes as in the proof of Theorem 3.1. (In order to conform with notations in the proof of Theorem 3.1, take a neighborhood W of p_0 defined by $|w^i| < \varepsilon/2$ in terms of a local coordinate system $w^1, \ldots, w^n$ in U as in the proof of Theorem 3.1. By taking a subsequence, we may also assume that $f(\gamma_k) \subset W$ for all k. We then proceed as in the proof of Theorem 3.1.)   QED.

If we set $Y = P_1(\mathbf{C})$ and $M = P_1(\mathbf{C}) - \{3 \text{ points}\}$, then conditions (1), (2), and (3) are met and we obtain the classical big Picard theorem.

Example 1.   Let $Y = P_2(\mathbf{C})$ and $p_j, j = 1, 2, 3, 4$, be four points in general position in $P_2(\mathbf{C})$. Drawing a complex line through each pair of these four points, we obtain a complete quadrilateral $Q = \bigcup_{i=0}^{5} L_i$ as in Example 1 of § 4 of Chapter IV. Let $M = P_2(\mathbf{C}) - Q$. Then conditions (1), (2), and (3) are satisfied. While (1) was verified in § 4 of Chapter IV, (2) is evident. To verify (3), let p be a point of $Q = \overline{M} - M$. Without loss of generality, we may assume that p is not on the line L_0, which we shall consider as the line at infinity so that $P_2(\mathbf{C}) - L_0 = \mathbf{C}^2$. By changing indices if necessary, we may assume that L_1 and L_2 meet at infinity, i.e., are parallel in $\mathbf{C}^2$ and that L_3 and L_4 are parallel in $\mathbf{C}^2 = P_2(\mathbf{C}) - L_0$. After a suitable linear transformation, we may assume that in $\mathbf{C}^2$

$$L_1 = \{(0, w) \in \mathbf{C}^2; \, w \in \mathbf{C}\}, \qquad L_2 = \{(1, w) \in \mathbf{C}^2; \, w \in \mathbf{C}\},$$
$$L_3 = \{(z, 0) \in \mathbf{C}^2; \, z \in \mathbf{C}\}, \qquad L_4 = \{(z, 1) \in \mathbf{C}^2; \, z \in \mathbf{C}\},$$

so that $N = \mathbf{C}^2 - (L_1 \cup L_2 \cup L_3 \cup L_4) = (\mathbf{C} - \{0, 1\}) \times (\mathbf{C} - \{0, 1\})$.
Let U be a neighborhood of p in $P_2(\mathbf{C})$. We may assume that U does
not meet L_0, i.e., U is contained in $\mathbf{C}^2 = P_2(\mathbf{C}) - L_0$. It is now clear
that there exists a neighborhood V of p in $\mathbf{C}^2$ such that $\bar{V} \subset U$ and the
distance between $N \cap (\mathbf{C}^2 - U)$ and $N \cap V$ is positive with respect to
d_N. Since $M = N - L_5$, the distance between $M \cap [P_2(\mathbf{C}) - U] =
M \cap (\mathbf{C}^2 - U)$ and $M \cap V$ is positive with respect to d_M. This proves
our assertion. From Theorem 6.1 we may conclude that *every holomorphic mapping f from the punctured disk D* into M = $P_2(\mathbf{C})$ − Q can
be extended to a holomorphic mapping from the disk D into $P_2(\mathbf{C})$, where Q
is a complete quadrilateral.*

In Example 1 of § 4 of Chapter IV,we constructed a complete hyperbolic manifold M by deleting a certain five lines from $P_2(\mathbf{C})$. It is not
clear whether Theorem 6.1 is valid for such a manifold M.

THEOREM 6.2.   *Assume M and Y satisfy conditions* (1), (2), *and* (3)
of Theorem 6.1. *Let X be a complex manifold of dimension m and A a
locally closed complex submanifold of dimension* $\leq m - 1$. *Then every
holomorphic mapping* $f : X - A \to M$ *can be extended to a holomorphic
mapping from X into Y.*

I do not know whether the theorem is valid when A is an analytic
subset with singular points.

Proof.   We may assume that $X = D^m = D \times D^{m-1}$ and $A = \{0\}
\times D^{m-1}$ so that $X - A = D^* \times D^{m-1}$. We denote a point $(z, t^1, \dots,
t^{m-1}) \in D^m$ by (z, t). Given a holomorphic mapping $f : D^* \times D^{m-1} \to M$,
the restriction of f to the punctured disk $D^* \times \{t\}$ can be extended to a
holomorphic mapping from the disk $D \times \{t\}$ into Y for each fixed t (by
Theorem 6.1). We have to show that this extended mapping $f : D^m \to Y$
is continuous at every point $(0, t)$ of A. It suffices of course to show that
f is continuous at the origin $(0, 0) \in D \times D^{m-1}$. Let $p = f(0, 0) \in Y$. If
p is in M, then the proof is the same as in the proof of Theorem 4.1.
We assume therefore that $p \in \bar{M} - M$. Let U be a neighborhood of p
in Y defined by $|w^i| < a$, $i = 1, \dots, n$, with respect to a local coordi-
nate system $w^1, \dots, w^n$ around p. Let V be a neighborhood of p descri-
bed in condition (3). Let b be the distance between $M \cap (Y - U)$ and
$M \cap V$ with respect to d_M. Let r be a positive number such that the disk
$\{(z, 0); |z| < r\}$ is mapped into V by f. Let r' be a positive number

such that

$$d_{D^{m-1}}(0, t) < b$$

if

$$|t^i| < r' \quad \text{for} \quad i = 1, \ldots, m-1.$$

If $0 < |z| < r$ and $|t^i| < r'$ for $i = 1, \ldots, m-1$, then

$$d_M(f(z, 0), f(z, t)) \leq d_{D^* \times D^{m-1}}((z, 0), (z, t)) \leq d_{D^{m-1}}(0, t) < r',$$

where the first inequality follows from the distance-decreasing property of $f : D^* \times D^{m-1} \to M$ and the second inequality follows from the distance decreasing property of the injection $t \in D^{m-1} \to (z, t) \in D^* \times D^{m-1}$. Since $f(z, 0)$ is in V and $f(z, t)$ is less than r' away from $f(z, 0)$, it follows that $f(z, t)$ is in U. By the Riemann extension theorem f is holomorphic in $\{(z, t); |z| < r \text{ and } |t^i| < r'\}$.   QED.

Example 2.   Let M be the quotient of a symmetric bounded domain by an arithmetically defined discrete group Γ. Let Y be the Satake compactification of M. If the action of Γ is free, then M and Y satisfy conditions (1), (2), (3) of Theorem 6.1. Even if the action is not free, M and Y satisfy similar conditions (see § 6 of Chapter VII). Theorem 6.2 applied to this example yields a simple proof of a result of Borel (see Kobayashi and Ochiai [1]). Although Y is a complex space with singularities, we shall see in Chapter VII that all the results in this section hold when M and Y are complex spaces.

VII

Generalization to Complex Spaces

1. Complex Spaces

We shall review quickly the definition and basic properties of complex spaces. For details and proofs, we refer the reader to Gunning and Rossi [1] and Narasimhan [1].

Let Ω be an open set in $\mathbf{C}^n$. A subset U of Ω is called an *analytic set* in Ω if every point $a \in \Omega$ has a neighborhood N_a such that $U \cap N_a$ is given as the common zeros of functions $f_1, \ldots, f_p$ holomorphic in N_a, i.e.,

$$U \cap N_a = \{x \in N_a; f_1(x) = \cdots = f_p(x) = 0\}.$$

It follows that U is closed in Ω, that $\Omega - U$ is dense in Ω if $U \neq \Omega$ and that $\Omega - U$ is connected if Ω is connected.

Let Ω and Ω' be two open neighborhoods of a point $a \in \mathbf{C}^n$. Let U and U' be analytic sets in Ω and Ω', respectively. We write $(\Omega, U) \sim (\Omega', U')$ at a if there exists an open neighborhood $\Omega'' \subset \Omega \cap \Omega'$ of the point a such that $U \cap \Omega'' = U' \cap \Omega''$. Clearly, $\sim$ is an equivalence relation. We call an equivalence class an *analytic germ* at a. We denote the analytic germ defined by (Ω, U) at a by U_a.

For each point $a \in \mathbf{C}^n$, let $\mathscr{O}_{n,a}$ denote the ring of germs of holomorphic functions at a. We denote by $\mathscr{O}_n$ the sheaf of germs of holomorphic functions on $\mathbf{C}^n$.

Let U be an analytic set in $\Omega \subset \mathbf{C}^n$. At each point $a \in \Omega$, we denote by $\mathscr{I}_a = \mathscr{I}(U_a)$ the set of germs of holomorphic functions in Ω vanishing on the germ U_a. Then $\mathscr{I}_a$ is an ideal of $\mathscr{O}_{n,a}$. If $a \notin U$, then $\mathscr{I}_a = \mathscr{O}_{n,a}$.

We denote by $\mathscr{I} = \mathscr{I}(U)$ the sheaf of ideals $\mathscr{I}_a$, $a \in \Omega$, of the analytic set U, i.e., $\mathscr{I} = \bigcup_{a \in \Omega} \mathscr{I}_a$. It is a subsheaf of $\mathscr{O}(\Omega) = \mathscr{O}_n \mid \Omega$.

Consider the quotient sheaf $\mathscr{O}(\Omega)/\mathscr{O}(U)$. Its stalk at $a \in \Omega$ is given by $\mathscr{O}_{n,a}/\mathscr{I}_a$; in particular, it is zero for $a \notin U$. The restriction of the quotient sheaf $\mathscr{O}(\Omega)/\mathscr{I}(U)$ to U, denoted by $\mathscr{O}_U$, is called the *sheaf of germs of holomorphic functions on U*. A section of the sheaf $\mathscr{O}_U$ over an open subset V of U is called a *holomorphic function* on V. This is equivalent to the following more elementary definition. A continuous, complex-valued function on V is said to be *holomorphic* if it is locally the restriction of a function holomorphic in Ω.

An analytic germ U_a at $a \in \Omega$ is said to be *reducible* if it is a union of two analytic germs at a, each of which is different from U_a. Then an analytic germ U_a is irreducible if and only if $\mathscr{I}(U_a)$ is a prime ideal of $\mathscr{O}_{n,a}$. Every analytic germ U_a can be uniquely written (up to order) as a finite union $U_a = \bigcup_{\nu=1}^k U_{\nu,a}$ of irreducible analytic germs $U_{\nu,a}$ such that, for each ν, $U_{\nu,a} \not\subset \bigcup_{\mu \neq \nu} U_{\mu,a}$.

For an analytic set U in $\Omega \subset \mathbf{C}^n$, the *local dimension* $\dim_a U$ at $a \in \Omega$ is defined as follows. If $a \notin U$, then $\dim_a U = -1$. If $a \in U$, then $\dim_a U = n - s$, where $s = \operatorname{codim}_a U$ is the dimension of a maximal dimensional linear subspace $L \subset \mathbf{C}^n$ through a such that a is an isolated point of $L \cap U$. We define $\dim U = \operatorname{Max}_{a \in U} \dim_a U$. If $\dim_a U = k$ for all $a \in U$, then U is said to be *pure k-dimensional*. If U is irreducible, then U is pure k-dimensional for some k.

Let U be an analytic subset in $\Omega \subset \mathbf{C}^n$. A point $a \in U$ is called a *regular* point of U of dimension k if a has neighborhood N in Ω such that $U \cap N$ is a k-dimensional submanifold of N. A point $a \in U$ is said to be *singular* if it is not regular. The set of singular points of U forms an analytic subset S of Ω such that $\dim S \leq \dim U - 1$.

A local geometric description of an analytic set is given as follows. Let U be an irreducible analytic set of dimension k in an open set $\Omega \subset \mathbf{C}^n$. Let $a \in U$. Then there exists a local coordinate system $z^1, \ldots, z^n$ in a neighborhood $D^n = \{\mid z^i \mid < 1; i = 1, \ldots, n\}$ of a in Ω with the following properties:

(1) The point a is at the origin $(0, \ldots, 0)$ and is an isolated point of the subset $\{(0, \ldots, 0, z^{k+1}, \ldots, z^n) \in U \cap D^n\}$ of $U \cap D^n$;

(2) If we set $D^k = \{(z^1, \ldots, z^k); \mid z^i \mid < 1 \text{ for } i = 1, \ldots, k\}$, the map $\pi : U \cap D^n \to D^k$ defined by

$$\pi(z^1, \ldots, z^k, z^{k+1}, \ldots, z^n) = (z^1, \ldots, z^k)$$

is surjective and proper;

(3) There exists an analytic subset S of dimension $\leq k - 1$ of $U \cap D^n$ such that S contains all singular points of $U \cap D^n$ and $\pi : U \cap D^n - S \to D^k - \pi(S)$ is a finitely sheeted covering space. [Moreover, $\pi(S)$ is an analytic subset of D^k of dimension $\leq k - 1$.]

Let X be a Hausdorff topological space and $\mathcal{O} = \mathcal{O}_X$ a subsheaf of the sheaf of germs of continuous functions on X. The pair $(X, \mathcal{O})$ is called a *complex space* if every point $a \in X$ has an open neighborhood U such that U is an analytic set in an open set $\Omega \subset \mathbf{C}^n$ and $\mathcal{O} \mid U$ is the sheaf of germs of holomorphic functions on the analytic set U. The sheaf $\mathcal{O}$ is called the *structure sheaf* of X. For any open subset U of X, the continuous sections of $\mathcal{O}$ over U are by definition the *holomorphic functions* on U. For the sake of simplicity, we often denote $(X, \mathcal{O})$ by X.

A continuous map $f : X \to Y$ of one complex space X into another, Y, is said to be *holomorphic* if $f^*(\mathcal{O}_{Y,f(a)}) \subset \mathcal{O}_{X,a}$ for every $a \in X$, i.e., if $h \circ f$ is a function holomorphic in a neighborhood of $a \in X$ whenever h is a function holomorphic in a neighborhood of $f(a) \in Y$.

Since a complex space is locally an analytic space, such local concepts as "regular point," "singular point," and "local dimension" can be defined in an obvious manner. A complex space X is *reducible* if it can be written as a union of two complex spaces, each of which is different from X. For a complex space X, its *dimension* dim X is defined to be the maximum of its local dimension. If its local dimension is k everywhere, X is said to be *pure k-dimensional*. If X is irreducible, its local dimension is a constant.

2. *Invariant Distances for Complex Spaces*

Let X be a connected complex space. Then we can define the Carathéodory pseudodistance c_X and the invariant pseudodistance d_X as in Chapter IV. In defining d_X, we have to use the fact that any two points p and q of X can be connected by a chain of analytic disks. To establish this fact, it suffices to prove that, given a (singular) point a of an irreducible analytic set U, there is a holomorphic mapping f of the unit disk D into U such that $f(D)$ contains a and also a regular point of U. Assuming that U is a k-dimensional analytic subset in an open set $\Omega \subset \mathbf{C}^n$, we take a local coordinate system $z^1, \ldots, z^n$ in Ω satisfying the three conditions (1), (2), and (3) of § 1. Since $\pi(S)$ is an analytic subset of dimension $\leq k - 1$ in D^k, we may assume that the coordinate system satisfies the following condition:

$$\pi(S) \cap D^1 = \{0\}, \qquad \text{where} \quad D^1 = (z^1, 0, \ldots, 0) \subset D^k.$$

By (1) we may assume, by taking a smaller neighborhood of a if necessary, that a is the only point of U which projects upon the origin 0 of D^k. By (3), $\pi : \pi^{-1}(D^1) - \{a\} \to D^1 - \{0\}$ is an s-sheeted covering projection, where s is a positive integer. Then it is not hard to see that there is a holomorphic mapping $f : D \to \pi^{-1}(D^1) \subset U$ such that

$$\pi[f(w)] = (w^s, 0, \ldots, 0) \in D^1 \subset D^k.$$

Then $f(0) = a$ and $f(w)$ is a regular point for $w \in D - \{0\}$.

A complex space X is said to be *hyperbolic* if d_X is a distance. A hyperbolic space X is said to be *complete* if it is complete with respect to d_X.

All the results in Chapter IV can be immediately generalized to complex spaces except those which make sense only for nonsingular complex manifolds (e.g., Theorems 3.4, 3.5, and 4.11).

Similarly, the results in Chapter V can be also generalized to complex spaces. For some results in §§ 2 and 3 of Chapter V, the following fact (see, for instance, Gunning and Rossi [1, p. 158]) plays an essential role. The space of holomorphic functions on a complex space is complete in the topology of uniform convergence on compact subsets.

For a complex space X, the tangent space $T_x(X)$ at $x \in X$ can be defined as the space of derivations on the ring of germs of holomorphic functions at x (see, for example, Gunning and Rossi [1, p. 152]). It is then not difficult to generalize Theorem 3.3 of Chapter V to a complex space.

3. *Extension of Mappings into Hyperbolic Spaces*

Theorem 3.1 of Chapter VI may be generalized as follows:

THEOREM 3.1. *Let Y be a hyperbolic complex space and D^* the punctured unit disk. Let $f : D^* \to Y$ be a holomorphic mapping such that, for a suitable sequence of points $z_k \in D^*$ converging to the origin, $f(z_k)$ converges to a point $y_0 \in Y$. Then f extends to a holomorphic mapping of the unit disk D into Y.*

COROLLARY 3.2. *If Y is a compact hyperbolic complex space, then every holomorphic mapping $f : D^* \to Y$ extends to a holomorphic mapping of D into Y.*

Proof. Let U be a neighborhood of y_0 in Y which is equivalent to an analytic subset in an open set Ω of $\mathbf{C}^n$. Let $w^1, \ldots, w^n$ be the coordinate system in $\mathbf{C}^n$. We may assume that y_0 is the origin of this coordinate system. Then the rest of the proof goes in the same way as in Theorem 3.1 of Chapter VI. QED.

Theorem 4.1 and Theorem 5.2 of Chapter VI may be generalized as follows without any change in their proofs.

THEOREM 3.3. *Let Y be a compact hyperbolic complex space. Let X be a complex manifold of dimension m and A an analytic subset of X of dimension $\leq m - 1$. Then every holomorphic mapping $f : X - A \to Y$ can be extended to a holomorphic mapping from X into Y.*

THEOREM 3.4. *Let Y be a complete hyperbolic complex space. Let X be a complex manifold of dimension m, and let A be a subset which is nowhere dense in an analytic subset, say B, of X with $\dim B \leq m - 1$. Then every holomorphic mapping $f : X - A \to Y$ can be extended to a holomorphic mapping of X into Y.*

COROLLARY 3.5. *Let Y be a complete hyperbolic complex space. Let X be a complex manifold of dimension m, and let A be an analytic subset of dimension $\leq m - 2$. Then every holomorphic mapping $f : X - A \to Y$ can be extended to a holomorphic mapping of X into Y.*

Similarly, Theorems 6.1 and 6.2 of Chapter VI may be generalized as follows:

THEOREM 3.6. *Let Z be a complex space and Y a complex subspace of Z satisfying the following three conditions:*

(1) *Y is hyperbolic;*
(2) *The closure of Y in Z is compact;*
(3) *Given a point p of $\bar{Y} - Y$ and a neighborhood U of p in Z, there exists a neighborhood V of p in Z such that $\bar{V} \subset U$ and the distance between $Y \cap (Z - U)$ and $Y \cap V$ with respect to d_Y is positive.*

Then every holomorphic mapping f from the punctured disk D^* into Y can be extended to a holomorphic mapping from the whole disk D into Z.

COROLLARY 3.7.   *Let Y and Z be as above. Let X be a complex manifold of dimension m and A a locally closed complex submanifold of dimension $\leq m - 1$. Then every holomorphic mapping $f : X - A \to Y$ can be extended to a holomorphic mapping from X into Z.*

We conclude this section by showing that, in theorems of the kind discussed here, the assumption that X is a nonsingular manifold is essential. Let Y be a projective algebraic variety in $P_n(\mathbf{C})$. Let $C(Y)$ be the affine cone of Y, i.e., the union of all complex lines through the origin of $\mathbf{C}^{n+1}$ representing the points of Y. We recall that Y is said to be projectively normal if $C(V)$ is a normal complex space. We make use of the following theorem (see Lang [1, p. 143]). *If Y is a normal projective algebraic variety, then Y can be imbedded into some $P_n(\mathbf{C})$ in such a way that Y is projectively normal, i.e., $C(Y)$ is normal.* In particular, let Y be a nonsingular projective algebraic manifold which is hyperbolic, e.g., a compact Riemann surface of genus >1. Then Y is projectively normal in some $P_n(\mathbf{C})$ and $C(Y)$ is nonsingular except at the origin. Let $\pi : C(Y) - \{0\} \to Y$ be the restriction of the natural projection $\mathbf{C}^{n+1} - \{0\} \to P_n(\mathbf{C})$. It is clear that π cannot be extended to a holomorphic mapping from $C(Y)$ into Y. This shows that Theorem 3.3 does not hold for $X = C(Y)$ and $A = \{0\}$ and hence that the cone $C(Y)$ is singular at the origin.

We give another example due to D. Eisenman. Let M be a compact complex manifold and L a negative line bundle over M. According to Grauert [2], the space L/M obtained from L by collapsing the zero section to a point is a complex space. Let f be the projection from $L - \{\text{zero section}\}$ onto M. It is clear that f cannot be extended to a (continuous) mapping from L/M into M. To obtain a counter-example, all we have to do is to take any compact hyperbolic manifold, e.g., a compact Riemann surface of genus ≥ 2 as M. This example shows also that if L is a negative line bundle over a compact hyperbolic manifold, then the point of L/M corresponding to the zero section of L is a singular point. For if it were a nonsingular point, f would be extended to a mapping from L/M into M by Theorem 4.1 of Chapter VI. More generally, we may take a complex vector bundle E which is negative in a certain sense in place of a negative line bundle (see Grauert [2]).

A similar reasoning shows that if we obtain a complex space by collapsing a complex subspace of a hyperbolic manifold to a point, then the resulting space has a singular point. This fact will be taken up in Chapter VIII.

4. *Normalization of Hyperbolic Complex Spaces*

A complex space X is said to be *normal* at a point $a \in X$ if the ring $\mathcal{O}_{X,a}$ of germs of holomorphic functions at a is integrally closed in its ring of quotients. If X is normal at every point of X, then X is said to be *normal*. A *normalization* of a complex space X is a pair (Y, π) consisting of a normal complex space Y and a surjective holomorphic mapping $\pi : Y \rightarrow X$ such that

   (i) $\pi : Y \rightarrow X$ is proper and $\pi^{-1}(a)$ is finite for every $a \in X$;

   (ii) If S is the set of singular points of X, then $Y - \pi^{-1}(S)$ is dense in Y and $\pi : Y - \pi^{-1}(S) \rightarrow X - S$ is biholomorphic.

The normalization theorem of Oka (see Oka [2], Narasimhan [1, p. 118]) says that every complex space X has a unique (up to an isomorphism) normalization (Y, π).

The following result is due to Kwack [1].

THEOREM 4.1.   *Let X and Y be complex spaces and let $f : Y \rightarrow X$ be a proper holomorphic mapping such that $f^{-1}(a)$ is finite for every $a \in X$. If X is hyperbolic, so is Y.*

COROLLARY 4.2.   *If (Y, π) is a normalization of a hyperbolic complex space X, then Y is also hyperbolic.*

Proof.   Since f is distance-decreasing, we have

$$d_Y(p, q) > 0 \quad \text{if} \quad f(p) \neq f(q).$$

Let p and q be two distinct points of Y such that $f(p) = f(q)$. Let V and W be disjoint open neighborhoods of p and q respectively such that

$$f^{-1}[f(p)] = (V \cup W) \cap f^{-1}[f(p)].$$

Assuming $d_Y(p, q) = 0$, we shall obtain a contradiction. Let $\{\gamma_n\}$ be a sequence of curves in Y joining p and q such that their lengths $L_Y(\gamma_n)$ measured in terms of d_Y satisfy

$$L_Y(\gamma_n) < \frac{1}{n}.$$

Let U_n be the closed ball of radius $1/n$ around the point $f(p)$ in X with respect to d_X. Since f is distance-decreasing, $f(\gamma_n)$ is contained in U_k for $k \leq n$. In other words,

$$\gamma_n \subset f^{-1}(U_k) \qquad \text{if} \quad k \leq n.$$

Since V and W are disjoint and γ_n is a curve from $p \in V$ to $q \in W$, there exists a point p_n on γ_n which is not in $V \cup W$. Since U_k is compact for sufficiently large k, so is $f^{-1}(U_k)$. Since $p_n \in f^{-1}(U_k)$ for $k \leq n$, it follows that the sequence $\{p_n\}$ has a subsequence which converges to a point, say p_0, of Y. Since $p_n \notin (V \cup W)$ and $V \cup W$ is open, the point p_0 is not in $V \cup W$. On the other hand, $f(p_n) \in U_n$ and $\cap \, U_n = \{f(p)\}$. Hence, $f(p_0) = f(p)$. This implies $p_0 \in f^{-1}[f(p)] = (V \cup W) \cap f^{-1}[f(p)]$ $\subset V \cup W$, contradicting the statement above that $p_0 \notin V \cup W$.   QED.

5. Complex V-Manifolds

Let G be a finite group of linear transformations of $\mathbf{C}^n$. Then the quotient space $\mathbf{C}^n/G$ is, algebraically, an affine algebraic variety and, analytically, a normal complex space (see H. Cartan [4]). A complex space X is called an n-dimensional *complex V-manifold* if every point p of X has a neighborhood U which is biholomorphic to a neighborhood of the origin in $\mathbf{C}^n/G_p$, where G_p is a finite group of linear transformations of $\mathbf{C}^n$ (which depends on p). It follows from the result of Cartan that a complex V-manifold is a normal complex space. The notion of V-manifold was introduced by Satake [1, 2] and has been extensively investigated by Baily [1, 2].

Let M be a complex manifold and G a finite group of holomorphic transformations of M leaving a point o fixed. With respect to a suitable local coordinate system with origin at o, the action of G is linear. In fact, let B be the open ball of radius r around o with respect to a Riemannian metric invariant by G, where r is chosen so small that B may be considered as a bounded domain in $\mathbf{C}^n$. Then G may be considered as a group of holomorphic transformations of a bounded domain B leaving a point o fixed. By a classical theorem of H. Cartan [3] (see also Bochner and Martin [1]), the action of G is linear with respect to a suitable local coordinate system.

Let M be a complex manifold and Γ a properly discontinuous group of holomorphic transformations of M. Since the isotropy subgroup of Γ

at each point of M is finite, the quotient space M/Γ is a complex V-manifold. In particular, if M is a hyperbolic manifold, e.g., a bounded domain in $\mathbf{C}^n$, and Γ is a discrete subgroup of the group $H(M)$ of holomorphic transformations of M, then M/Γ is a complex V-manifold.

In §§ 4 and 5 of Chapter VI, we discussed the problem of extending a holomorphic mapping $f : X - A \to M$ to a holomorphic mapping from X into M, where X is a complex manifold, A is an analytic subset of X, and M is a hyperbolic manifold. As I have indicated in § 3, the results of §§ 4 and 5 of Chapter VI can be generalized to the case where M is a hyperbolic complex space. On the other hand, the assumption that X is a nonsingular complex manifold seems to be rather essential. It is clear that the results can be generalized to the case where X is a complex V-manifold.

6. *Invariant Distances on* M/Γ

Let M be a complex manifold (or more generally, a complex space) and Γ a properly discontinuous group of holomorphic transformations of M. We set $Y = M/\Gamma$. In addition to the pseudodistance d_Y defined earlier, we can define another pseudodistance d_Y' as follows. Let $\pi : M \to Y = M/\Gamma$ be the natural projection. A holomorphic mapping f from a complex space X into Y is said to be *liftable* if there exists a holomorphic mapping $\tilde{f} : X \to M$ such that $f = \pi \circ \tilde{f}$. It is said to be *locally liftable* if every point of X has a nieghborhood U such that the restriction of f to U is liftable. If X is simply connected, then a locally liftable mapping $f : X \to Y$ is liftable. If the group Γ acts freely on M so that M is a covering space of Y, then every holomorphic mapping $f : X \to Y$ is locally liftable. In the definition of d_Y, we used chains of holomorphic mappings from the unit disk D into Y. If we use only locally liftable holomorphic mappings from D into Y, we obtain a new pseudodistance which will be denoted by d_Y'. Since D is simply connected, to define d_Y' it suffices to consider only liftable holomorphic mappings from D into Y. In general, we have

$$d_Y'(p, q) \geqq d_Y(p, q) \qquad \text{for} \quad p, q \in Y.$$

If Γ acts freely on M so that $\pi : M \to Y$ is a covering projection, then $d_Y' = d_Y$. The following proposition is trivial.

PROPOSITION 6.1.   *Let M be a complex space and Γ a properly discontinuous group of holomorphic transformations of M. Let f be a locally liftable holomorphic mapping from a complex space X into $Y = M/\Gamma$. Then*

$$d_{Y'}(f(p), f(q)) \leqq d_X(p, q) \qquad for \quad p, q \in X.$$

The following proposition is a generalization of Proposition 1.6 of Chapter IV. The proof is similar.

PROPOSITION 6.2.   *Let M be a complex space and Γ a properly discontinuous group of holomorphic transformations of M. Let $\pi : M \to Y = M/\Gamma$ be the natural projection. Let $p, q \in Y$ and $\tilde{p}, \tilde{q} \in M$ such that $\pi(\tilde{p}) = p$ and $\pi(\tilde{q}) = q$. Then*

$$d_{Y'}(p, q) = \inf_{\tilde{q}} d_M(\tilde{p}, \tilde{q}),$$

where the infimum is taken over all $\tilde{q} \in M$ such that $\pi(\tilde{q}) = q$.

The proof of the following proposition is similar to that of Theorem 4.7 of Chapter IV.

PROPOSITION 6.3.   *Let M, Γ and Y be as above. Then $d_{Y'}$ is a (complete) distance if and only if M is a (complete) hyperbolic space, i.e., if and only if d_M is a (complete) distance.*

Many of the results in Chapters IV, V, and VI may be generalized to M/Γ and locally liftable holomorphic mappings. We mention one.

THEOREM 6.4.   *Let M be a hyperbolic complex space and Γ a properly discontinuous group of holomorphic transformations. Let f be a locally liftable holomorphic mapping from the punctured disk D^* into $Y = M/\Gamma$ such that, for a suitable sequence of points $z_k \in D^*$ converging to the origin, $f(z_k)$ converges to a point $p_0 \in Y$. Then f extends to a holomorphic mapping from the unit disk D into Y.*

VIII

Hyperbolic Manifolds and Minimal Models

1. Meromorphic Mappings

There are several definitions of a meromorphic mapping from a complex space into another complex space. Remmert [1] and Stoll [2] have made systematic studies on meromorphic mappings.

A *meromorphic mapping* f from a complex space X into a complex space Y in the sense of Remmert is a correspondence satisfying the following conditions:

(1) For each point x of X, $f(x)$ is a nonempty compact subset of Y;

(2) The graph $\Gamma_f = \{(x, y) \in X \times Y; \ y \in f(x)\}$ of f is a connected complex subspace of $X \times Y$ with $\dim \Gamma_f = \dim X$;

(3) There exists a dense subset X^* of X such that $f(x)$ is a single point for each x in X^*.

Denote by $\tilde{\pi}$ the projection from $X \times Y$ onto X. Denote by π the restriction of $\tilde{\pi}$ to the graph Γ_f. Since $f(x) = \pi^{-1}(x) = \Gamma_f \cap \tilde{\pi}^{-1}(x)$, it follows that $f(x)$ is a complex subspace of Y.

Let E be the set of points of Γ_f where π is degenerate, i.e.,

$$E = \{(x, y) \in \Gamma_f; \ \dim f(x) > 0\}.$$

Let

$$N = \pi(E) = \{x \in X; \ \dim f(x) > 0\}.$$

Then E is a closed complex subspace of codimension ≥ 1 of Γ_f and N

105

is a closed complex subspace of codimension $\geqq 2$ of X (see Remmert [1]). It is then clear that $f : X - N \to Y$ is a holomorphic mapping.

We shall consider the case where Y is a complex projective space. We start with the following two general propositions.

PROPOSITION 1.1.   *Let X be a normal complex space and A a locally closed complex subspace of codimension $\geqq 2$. Every complex vector bundle ξ on $X - A$ can be (uniquely) extended to a complex vector bundle $\tilde{\xi}$ over X. Every holomorphic cross section of $\tilde{\xi}$ over $X - A$ can be extended to a holomorphic cross section over X.*

Proof.   Let m be the fiber dimension of ξ. Let $\{U_\alpha\}$ be an open cover of X. Let $\{g_{\alpha\beta}\}$ be the transition functions for the bundle ξ so that each

$$g_{\alpha\beta} : U_\alpha \cap U_\beta - A \to GL(m; \mathbf{C})$$

is holomorphic. Extend each entry of the matrix $g_{\alpha\beta}$ to a holomorphic function on $U_\alpha \cap U_\beta$. This is possible since $U_\alpha \cap U_\beta$ is normal and A has codimension $\geqq 2$. Then we obtain a holomorphic mapping $\tilde{g}_{\alpha\beta} : U_\alpha \cap U_\beta \to M(m; \mathbf{C})$, where $M(m; \mathbf{C})$ denotes the space of $m \times m$ complex matrices. Since $\tilde{g}_{\alpha\beta}\tilde{g}_{\beta\alpha} = 1$ on $U_\alpha \cap U_\beta - A$, it follows that $\tilde{g}_{\alpha\beta}\tilde{g}_{\beta\alpha} = 1$ on $U_\alpha \cap U_\beta$ and hence $\tilde{g}_{\alpha\beta}$ maps $U_\alpha \cap U_\beta$ into $GL(m; \mathbf{C})$. Let $\tilde{\xi}$ be the vector bundle over X defined by the transition functions $\{\tilde{g}_{\alpha\beta}\}$. The second assertion is trivial, since a holomorphic cross section is locally a system of m holomorphic functions.   QED.

Let $w^0, w^1, \ldots, w^m$ be a homogeneous coordinate system for the projective space $P_m(\mathbf{C})$. Let V_i be the open set in $P_m(\mathbf{C})$ defined by $w^i \neq 0$. We set $h_{ij} = w^j/w^i$ on $V_i \cap V_j$. Let η be the complex line bundle over $P_m(\mathbf{C})$ defined by the transition functions $\{h_{ij}\}$. This bundle admits $m + 1$ linearly independent holomorphic cross sections, say $s_0, s_1, \ldots, s_m$, which may be given as follows. Over each V_j, the bundle η is trivial and

$$s_k(p) = [p, w^k(p)/w^j(p)] \approx V_j \times \mathbf{C} \qquad \text{for} \quad p \in V_j.$$

These cross sections induce an imbedding of $P_m(\mathbf{C})$ into $P_m(\mathbf{C})$, which is nothing but the identity mapping. Let X be a normal complex space and A a locally closed complex subspace of codimension $\geqq 2$. Let $f : X - A \to P_m(\mathbf{C})$ be a holomorphic mapping. Let $\xi = f^{-1}\eta$; it is a complex line bundle over $X - A$. By Proposition 1.1, ξ may be extended to a complex line bundle $\tilde{\xi}$ over X. Let $f^*s_0, f^*s_1, \ldots, f^*s_m$ be the cross sections of ξ induced from $s_0, s_1, \ldots, s_m$. They induce precisely the mapping $f : X - A \to P_m(\mathbf{C})$. In other words, if f is given locally by

$$w^i = f^i(x), \qquad x \in X, \qquad \text{for} \quad i = 0, 1, \ldots, m,$$

then the meromorphic function f^i/f^j on $X - A$ coincides with f^*s_i/f^*s_j. Now, by the second statement in Proposition 1.1, the cross sections f^*s_0, $f^*s_1, \ldots, f^*s_m$ can be extended to holomorphic cross sections $t_0, t_1, \ldots,$ t_m of $\tilde{\xi}$ over X. If these sections have no common zeros on X, then they induce a holomorphic mapping of X into $P_m(\mathbf{C})$ which extends f. But in general they have common zeros. At any rate, the t_i/t_j are all meromorphic functions on X. Hence the meromorphic functions f^i/f^j on $X - A$ can be extended to meromorphic functions on X. (This last fact is of course a consequence of the general fact that every meromorphic function on $X - A$ can be extended to a meromorphic function on X if X is normal and A has codimension ≥ 2.) We have thus proved that *if X is a normal complex space and A is a locally closed complex subspace of codimension ≥ 2 and if $f : X - A \to P_m(\mathbf{C})$ is a holomorphic mapping, then every point of X has a neighborhood U and a holomorphic mapping $f_U : U \to \mathbf{C}^{m+1}$ such that f_U maps $U \cap (X - A)$ into $\mathbf{C}^{m+1} - \{0\}$ and induces the mapping f on $U \cap (X - A)$.* We shall show that f is meromorphic in the sense of Remmert, or more precisely, f is the restriction to $X - A$ of a meromorphic mapping of X into $P_m(\mathbf{C})$. In order to define the graph Γ_f, we consider the complex subspace T of $X \times P_m(\mathbf{C})$ defined locally as follows. If the mapping $f_U : U \to \mathbf{C}^{m+1}$ is given by

$$w^i = f^i(x), \qquad x \in U, \qquad \text{for} \quad i = 0, 1, \ldots, m,$$

then $T \cap [U \times P_m(\mathbf{C})]$ is by definition the common zeros of

$$f^i(x) w^j - f^j(x) w^i = 0 \qquad i, j = 0, 1, \ldots, m,$$

where $w^0, w^1, \ldots, w^m$ is considered as a homogeneous coordinate system of $P_m(\mathbf{C})$. Let Γ_f be the closure of $T \cap [(X - A) \times P_m(\mathbf{C})]$ and $f(x)$ $= \{y \in P_m(\mathbf{C}); (x, y) \in \Gamma_f\}$. It is then not difficult to check that f is a meromorphic mapping with graph Γ_f. (The space T is too large to be the graph of a meromorphic mapping, since the set $\{y \in P_m(\mathbf{C}); (x, y) \in T\}$

coincides with $P_m(\mathbf{C})$ if x is a point of U where all f^i's vanish.) See Remmert [1], Stoll [1], and Andreotti and Stoll [1] for more details.

What we have just seen for the case $Y = P_m(\mathbf{C})$ justifies Remmert's definition of meromorphic mappings.

From Corollary 3.5 of Chapter VII we obtain

THEOREM 1.2. *Let f be a meromorphic mapping from a complex manifold X into a complete hyperbolic space Y. Then f is holomorphic.*

Observe that X is assumed to be a nonsingular manifold.

Remark. If we make use of Theorem 3.1 of Chapter VI and the property of a meromorphic mapping that $f(x)$ is nonempty, then we can actually show that *every meromorphic mapping f from a complex manifold into a hyperbolic space* (complete or not) *is holomorphic.*

2. *Strong Minimality and Minimal Models*

We say that a complex space X is *strongly minimal* if every meromorphic mapping from a complex space U into X is holomorphic at every simple (i.e., nonsingular) point of U. This is the definition used by Weil [1, p. 27] in showing that every Abelian variety is minimal.

Let X and X' be complex spaces. We say that X and X' are *bimeromorphic* to each other if there exist meromorphic mappings $f : X \to X'$ and $g : X' \to X$ such that

$$x \in g[f(x)] \quad \text{for} \quad x \in X \quad \text{and} \quad x' \in f[g(x')] \quad \text{for} \quad x' \in X'.$$

We consider the case where $f : X \to X'$ is holomorphic and hence single-valued. Let $N' = \{x' \in X'; \ \dim g(x') > 0\}$ and $N = f^{-1}(N')$. Then, codim $N \geq 1$ and codim $N' \geq 2$ (see § 1). The restricted mapping $f : X - N \to X' - N'$ is biholomorphic. We call

$$g : (X', N') \to (X, N)$$

a *monoidal transformation* with center N' and its inverse

$$f : (X, N) \to (X', N')$$

a *contraction*. For details on monoidal transformations and contractions we refer the reader to Moisezon [1].

Given a class of bimeromorphically equivalent complex spaces, a space X_0 is called the *minimal model* of the class if, for every space X in the class, there is a contraction $f : (X, N) \to (X_0, N_0)$. It is clear that in the given class there is at most one minimal model.

It is clear that if X is a strongly minimal complex manifold, then it is the minimal model in its class of bimeromorphically equivalent complex manifolds.

Theorem 1.2 may be restated as follows:

THEOREM 2.1.   *Every complete hyperbolic space is strongly minimal.*

As we remarked at the end of § 1, we may drop the assumption of completeness. But we are primarily interested in compact spaces. This theorem generalizes a result of Igusa [1] (every compact Kaehler manifold with negative constant holomorphic sectional curvature is strongly minimal) and a result of Shioda [1] (a complex manifold which has a bounded domain of $\mathbf{C}^n$ as a covering manifold is strongly minimal).

The following result implies that every complex torus is also strongly minimal.

THEOREM 2.2.   *If X is a complex space which has a closed complex subspace $\tilde{X}$ of $\mathbf{C}^N$ as a covering space, then X is strongly minimal.*

Proof.   Let U be a complex manifold of dimension m. Let f be a meromorphic mapping of U into X. We want to prove that f is a holomorphic mapping from U into X. Since the problem is local with respect to the domain U, we may assume that U is a ball in $\mathbf{C}^m$ and that f is holomorphic on $U - A$ where A is an analytic subset of dimension $\leq m - 2$. Then $U - A$ is simply connected so that the holomorphic mapping $f : U - A \to X$ can be lifted to a holomorphic mapping $\tilde{f} : U - A \to \tilde{X} \subset \mathbf{C}^N$. Since $\tilde{f}$ is given by N holomorphic functions, it can be extended to a holomorphic mapping $\tilde{f} : U \to \mathbf{C}^N$ by Hartogs' theorem. Since $\tilde{X}$ is closed, $\tilde{f}$ maps U into $\tilde{X}$. If we denote by π the projection $\tilde{X} \to X$, then $\pi \circ f$ is a holomorphic mapping of U into X which coincides with f on $U - A$.   QED.

Theorem 2.1 implies that a compact Kaehler manifold with negative holomorphic sectional curvature is strongly minimal and hence is the minimal model in its class of bimeromorphically equivalent complex manifolds. We shall now show that a compact Kaehler manifold with negative Ricci tensor is the minimal model in its class of bimeromorphically equivalent complex manifolds.

Let E be a holomorphic vector bundle over a compact complex manifold M. Let $\Gamma(E)$ denote the space of holomorphic cross sections of E. If the restriction mapping $\sigma \in \Gamma(E) \to \sigma(x) \in E_x$ is surjective for each point x of M, then E is said to *have no base points*. Assuming that E has no base points, set

$$r = \dim E_x, \qquad k + r = \dim \Gamma(E).$$

To each point x of M, we assign the kernel of the restriction mapping $\Gamma(E) \to E_x$, which is a k-dimensional subspace of $\Gamma(E)$. In this way we obtain a holomorphic mapping of M into the complex Grassmann manifold $G_{k,r}(\mathbf{C})$ of k-planes in the $(k + r)$-dimensional vector space $\Gamma(E)$. If this mapping $M \to G_{k,r}(\mathbf{C})$ is an imbedding, then the bundle E is said to be *very ample*.

If a (local) holomorphic section of E is a covariant holomorphic tensor field of M, then E is called a *covariant* holomorphic vector bundle over M. The bundle of complex $(p, 0)$-forms on M is a covariant holomorphic vector bundle. In particular, the canonical line bundle, i.e., the bundle of $(n, 0)$-forms on M (where $n = \dim M$), is a covariant holomorphic vector bundle.

We prove first the following theorem.

THEOREM 2.3.   *Let E be a very ample covariant holomorphic vector bundle over a compact complex manifold M. If $f : M \to M$ is a meromorphic mapping which is nondegenerate at some point, then f is biholomorphic. In particular, every bimeromorphic mapping f of M onto M is biholomorphic.*

Proof.   Let $\omega \in \Gamma(E)$. Since ω is a covariant holomorphic tensor field on M, f induces a covariant holomorphic tensor field $f^*(\omega)$ on $M - N$, where N is an analytic subset of codimension ≥ 2 in M. Since $f^*(\omega)$ is given, locally, by a system of holomorphic functions, $f^*(\omega)$ extends to a holomorphic section of E over M by Hartogs' theorem. We shall show that the linear mapping $f^* : \Gamma(E) \to \Gamma(E)$ is injective. Let U and U' be open sets in M such that f maps U biholomorphically onto U'. If $\omega \in \Gamma(E)$ and $f^*(\omega) = 0$, then $f^*(\omega)$ vanishes identically on U and, consequently, ω vanishes identically on U'. Since ω is holomorphic, it vanishes identically on M. This proves our assertion. Since dim $\Gamma(E) < \infty$ by compactness of M, it follows that $f^* : \Gamma(E) \to \Gamma(E)$ is a linear automorphism. This linear isomorphism f^* induces an automorphism φ of the Grassmann manifold $G_{k,r}(\mathbf{C})$ of k-planes in $\Gamma(E)$, where $r = $ dim E_x and $k + r =$ dim $\Gamma(E)$ as above. We consider M as a complex submanifold of $G_{k,r}(\mathbf{C})$. Then $f : M \to M$ is the restriction of $\varphi^{-1} : G_{k,r}(\mathbf{C}) \to G_{k,r}(\mathbf{C})$ to M and hence is biholomorphic.   QED.

COROLLARY 2.4. *Let E be a very ample covariant holomorphic vector bundle over a compact complex manifold M. Then M is the minimal model in its class of bimeromorphically equivalent complex spaces.*

Proof. Let X be a complex space bimeromorphic to M. Let $f : X \to M$ and $g : M \to X$ be meromorphic mappings such that

$$x \in g[f(x)] \quad \text{for } x \in X \quad \text{and} \quad p \in f[g(p)] \quad \text{for } p \in M.$$

We want to show that f is holomorphic, i.e., single-valued. By Theorem 2.3, $f \circ g : M \to M$ is biholomorphic and $f[g(p)] = p$ for $p \in M$. Since $g : M \to X$ is surjective by $x \in g[f(x)]$, it follows that f is single-valued.
           QED.

Example 1. Let M be a compact Kaehler manifold with negative Ricci tensor or, more generally, with negative first Chern class $c_1(M)$. From a result of Kodaira [1] it follows that a suitable positive power K^m of the canonical line bundle K of M is very ample. (The converse is a triviality.) Theorem 2.3 implies that a meromorphic mapping of M into itself which is nondegenerate at some point is a biholomorphic mapping of M onto itself. This is a result of Peters [1]. By Corollary 2.4, M is the minimal model in its class of meromorphically equivalent complex spaces. This is due to P. Kiernan, who has pointed out to me that Theorem 2.3 implies Corollary 2.4 immediately.

Example 2. Consider r nonsingular hypersurfaces $V_1, \ldots, V_r$ of degree $d_1, \ldots, d_r$ in $P_{n+r}(\mathbf{C})$ respectively. Suppose that they are in general position so that the intersection $M = V_1 \cap \cdots \cap V_r$ is a nonsingular manifold of dimension n. If we denote by h the second cohomology class of M corresponding to a hyperplane section of M, then a formula on Chern classes (see Hirzebruch [1, p. 159]) implies $c_1(M) = (n + r + 1 - \sum_{i=1}^{r} d_i)h$. If $\sum d_i > n + r + 1$, then $c_1(M) < 0$ and

Theorem 2.3 and Corollary 2.4 apply to M. In particular, a nonsingular hypersurface M of degree d in $P_{n+1}(\mathbf{C})$ with $d > n + 2$ has $c_1(M) < 0$ and is the minimal model in its class of bimeromorphically equivalent complex spaces.

3. Relative Minimality

Let X and X' be complex spaces which are bimeromorphic to each other with meromorphic mappings $f : X \to X'$ and $g : X' \to X$ such that

$$x \in g[f(x)] \quad \text{for} \quad x \in X \quad \text{and} \quad x' \in f[g(x')] \quad \text{for} \quad x' \in X'.$$

In § 2, we defined the notions of monoidal transformation and contraction. If f is holomorphic, i.e., if $f : (X, N) \to (X', N')$ is a contraction, then we write

$$X > X'.$$

The minimal model X_0 in a class of bimeromorphically equivalent complex spaces satisfies, by definition, the inequality $X > X_0$ for every complex space X in the class. On the other hand, a space X in a class of bimeromorphically equivalent complex spaces is said to be *relatively minimal* if the class contains no space X' (other than X itself) such that $X > X'$. The minimal model may or may not exist but is unique (if it exists) in each class. On the other hand, relatively minimal models always exist but are not necessarily unique in each class of bimeromorphically equivalent complex manifolds (see Corollary 3.5 below). The following statement is obvious.

THEOREM 3.1.   *The minimal model is relatively minimal in its class of bimeromorphically equivalent complex spaces.*

We shall find other sufficient conditions for a complex manifold to be relatively minimal. For this purpose, we review quickly results of Kodaira and Spencer [1] on divisor class groups. A *divisor* of an n-dimensional compact complex manifold M is a finite sum of the form $\sum n_i V_i$, where $n_i \in \mathbf{Z}$ and V_i is an $(n-1)$-dimensional closed complex subspace of M. The set of divisors on M forms an additive group $G(M)$, called the *group of divisors*. Every meromorphic function f on M defines a divisor $(f) = (f)_0 - (f)_\infty$, where $(f)_0$ is the variety of zeros of f and $(f)_\infty$ is the variety of poles of f. The subgroup of $G(M)$ consisting of divisors (f) of meromorphic functions f is denoted by $G_l(M)$. The factor group $A(M) = G(M)/G_l(M)$ is called the *group of linear equivalence* on M. Each divisor $\sum n_i V_i$ of M gives rise to a complex line bundle over M as follows. Let $\{U_\alpha\}$ be an open cover of M, where each U_α is sufficiently small. Let $f_{\alpha i}$ be a holomorphic function on U_α such that $U_\alpha \cap V_i$ is defined by $f_{\alpha i} = 0$. If $U_\alpha \cap U_\beta \neq \emptyset$, we set

$$g_{\alpha\beta} = \prod_i (f_{\alpha i}/f_{\beta i})^{n_i}.$$

We associate to $\sum n_i V_i$ the line bundle defined by the transition functions $\{g_{\alpha\beta}\}$ and obtain a homomorphism of the group $G(M)$ into the group F of line bundles over M, whose kernel is exactly $G_l(M)$, and hence an isomorphism of $A(M) = G(M)/G_l(M)$ into F. It is known that if M is a closed complex submanifold of a complex projective space (i.e., if M is projective algebraic), then $A(M) \approx F$.

Given a complex line bundle $\xi \in F$ with transition functions $\{g_{\alpha\beta}\}$ we define a 2-cocycle $\{c_{\alpha\beta\gamma}\}$ of M by

$$\log g_{\alpha\beta} + \log g_{\beta\gamma} + \log g_{\gamma\alpha} = 2\pi i c_{\alpha\beta\gamma}, \qquad c_{\alpha\beta\gamma} \in Z.$$

This induces a homomorphism $F \to H^2(M; Z)$, denoted by c_1. The class $c_1(\xi) \in H^2(M; Z)$ is called the *first Chern class* of ξ. Let P be the kernel of $c_1 : F \to H^2(M; Z)$; it is called the *Picard variety* attached to M. In other words, P is the subgroup of F consisting of line bundles with trivial Chern class.

Let Ω_0 denote the sheaf of germs of holomorphic functions on M. We define a homomorphism from $H^1(M; \Omega_0)$ onto P as follows. Given a 1-cocycle $\{h_{\alpha\beta}\}$ representing an element of $H^1(M; \Omega^0)$, we set

$$g_{\alpha\beta} = \exp h_{\alpha\beta}.$$

Then $\{g_{\alpha\beta}\}$ defines an element of P. We have now an exact sequence

$$0 \to H^1(M; Z) \to H^1(M; \Omega^0) \to P \to 0,$$

where $H^1(M; Z) \to H^1(M; \Omega^0)$ is induced by $m \in Z \to 2\pi i m \in \Omega^0$.

Assume that M is a compact Kaehler manifold. Then the homomorphism $F \to H^0(M; Z)$ maps F onto $H^{1,1}(M; Z)$, where $H^{1,1}(M; Z)$ is the subgroup of $H^2(M; Z)$ consisting of elements which are mapped into $H^{1,1}(M; C)$ under the natural map $H^2(M; Z) \to H^2(M; C)$. Thus we

have
$$F/P \approx H^{1,1}(M; Z).$$

Since $H^1(M; \Omega^0) \approx H^{0,1}(M; \mathbf{C})$, we have
$$P \approx H^{0,1}(M; \mathbf{C})/J$$
with
$$J = \{\omega = 2\pi i \prod_{0,1} Hc; \ c \in H^1(M; Z)\},$$

where Hc denotes the harmonic form representing the class c and $\prod_{0,1}$ denotes the projection onto the space of $(0, 1)$-forms.

THEOREM 3.2. *Let M and M' be compact complex manifolds of dimension n, $F(M)$ and $F(M')$ the groups of line bundles over M and M' respectively, and $P(M)$ and $P(M')$ the Picard varieties attached to M and M' respectively. Let $f : (M, N) \to (M', N')$ be a contraction. Then $f^* : F(M')$ $\to F(M)$ is injective and maps $P(M')$ into $P(M)$, but the image $f^*F(M')$ does not contain the line bundle over M defined by the divisor mN for any nonzero integer m.*

Proof. We first note that N is given as the set of zeros of the Jacobian of f so that codim $N = 1$. As we stated in § 1, a result of Remmert implies that codim $N' \geq 2$. Let ξ' be a line bundle over M' and let $\xi = f^*\xi'$ be the induced line bundle over M. Assume that ξ is a trivial line bundle. Then ξ admits a holomorphic cross section σ which never vanishes on M. Since $f : M - N \to M' - N'$ is biholomorphic, f^* induces an isomorphism between $\xi' \mid (M' - N')$ and $\xi \mid (M - N)$. Let σ' be the holomorphic cross section of ξ' over $M' - N'$ corresponding to the section $\sigma \mid (M - N)$. Since codim $N' \geq 2$, we can extend σ' to a holomorphic section, denoted again by σ', over M' by Hartogs' theorem. Since the set of zeros of σ' on M' is either an empty set or an analytic subset of codimension 1 and must be contained in N', it must be empty. This shows that σ' never vanishes on M' and hence ξ' is trivial. The commutativity of the diagram

$$\begin{array}{ccccc}
0 \to & P(M) & \to & F(M) & \to & H^2(M; Z) \\
& \uparrow & & \uparrow & & \uparrow \\
0 \to & P(M') & \to & F(M') & \to & H^2(M'; Z)
\end{array}$$

implies that f^* maps $P(M')$ into $P(M)$.

To prove the last assertion of the theorem, let ξ be the line bundle associated to the divisor mN and let σ be a holomorphic cross section of ξ whose zero set is precisely mN (multiplicity counted). Assume that $\xi = f^*\xi'$, where ξ' is a line bundle over M'. Let σ' be the holomorphic cross section of ξ' over $M' - N'$ corresponding to $\sigma \mid (M - N)$. We extend σ' to a holomorphic cross section over M'. Again looking at the zeros of σ' as above, we obtain a contradiction unless $m = 0$.   QED.

COROLLARY 3.3. *If M is a compact complex manifold with its Picard variety $P(M) = 0$ and its second Betti number $b_2(M) = 0$, then M is relatively minimal in the class of compact complex manifolds which are bimeromorphic to M.*

Proof. Assuming the contrary, let $(M, N) \to (M', N')$ be a contraction. Since $P(M) = 0$, we have $P(M') = 0$. From

$$\text{rank } F(M) = \text{rank}[F(M)/P(M)] \leq b_2(M) = 0,$$

we obtain

$$\text{rank } F(M') = \text{rank } F(M) = 0.$$

On the other hand, the bundles associated to the divisors mN, $m \neq 0$, are nontrivial.   QED.

COROLLARY 3.4. *If M is a compact Kaehler manifold with $\dim H^{1,1}(M; \mathbf{C}) = 1$, then M is relatively minimal in the class of compact Kaehler manifolds which are bimeromorphic to M.*

Proof. The proof is similar to that of Corollary 3.3. Assuming the contrary, let $(M, N) \to (M', N')$ be a contraction, where M' is a compact Kaehler manifold. The following two facts imply the corollary immediately:

(1) $\text{rank}[F(M')/P(M')] = \text{rank } H^{1,1}(M'; Z) \geq 1$;

(2) $\dim P(M) = \dim H^{0,1}(M; \mathbf{C}) = \dim H^{1,0}(M; \mathbf{C})$
$$= \dim H^{1,0}(M'; \mathbf{C}) = \dim H^{0,1}(M'; \mathbf{C})$$
$$= \dim P(M').$$

Since (1) is a well-known property of a compact Kaehler manifold M', we prove only (2). Since M and M' are compact Kaehler manifolds, we have

$H^{1,0}(M; \mathbf{C}) = $ the space of holomorphic 1-forms on M,

$H^{1,0}(M'; \mathbf{C}) = $ the space of holomorphic 1-forms on M'.

It suffices to show that the map $f^* : H^{1,0}(M'; \mathbf{C}) \to H^{1,0}(M; \mathbf{C})$ is an isomorphism, where f is the contraction $(M, N) \to (M', N')$. Let ω' be a holomorphic 1-form on M'. If $f^*\omega' = 0$ on M, then $\omega' = 0$ on $M' - N'$ and hence $\omega' = 0$ on M' since codim $N' \geqq 2$. Let ω be a holomorphic 1-form on M. Consider the holomorphic 1-form on $M' - N'$ which corresponds to $\omega \mid (M - N)$ under f and then extend it to a holomorphic 1-form ω' on M'. Then $f^*\omega' = \omega$ on $M - N$ and hence $f^*\omega' = \omega$ on M.
                                                            QED.

COROLLARY 3.5.   *Let M be a compact (Kaehler) manifold. Then in the class of compact (Kaehler) manifolds bimeromorphic to M, there is one which is relatively minimal.*

Proof.   If M is not relatively minimal, let $(M, N) \to (M', N')$ be a contraction. By Theorem 3.2, rank $F(M) >$ rank $F(M')$. If M' is not relatively minimal, let $(M', N_1') \to (M'', N'')$ be a contraction so that rank $F(M') >$ rank $F(M'')$. Since rank $F(M)$ is finite, this process must stop eventually.   QED.

As we shall see shortly, a relative minimal model in the given class need not be unique, (see Example 2 below).

Example 1.   Let M be a compact Kaehler manifold with positive sectional curvature (or more generally, with positive holomorphic bi-sectional curvature). It is known (Bishop and Goldberg [1], Goldberg and Kobayashi [1]) that dim $H^{1,1}(M; \mathbf{C}) = 1$. By Corollary 3.4, M is relatively minimal in the class of compact Kaehler manifolds which are bimeromorphic to M.

Example 2.   Let M be a compact homogeneous Kaehler manifold of the form G/H, where G is a connected compact semisimple Lie group and H is a closed subgroup with 1-dimensional center. Every irreducible compact Hermitian symmetric space is such a manifold. From the exact homotopy sequence of the bundle $G \to G/H$, it follows that $b_2(M) = 1$. By Corollary 3.4, M is relatively minimal in the class of compact Kaehler manifolds which are bimeromorphic to M. According to Goto [1], $M = G/H$ is a rational algebraic manifold and hence is bimeromorphic to a projective space.

Example 3.   Let M be a nonsingular closed hypersurface in $P_{n+1}(\mathbf{C})$ with $n \geq 3$. More generally, let $M = V_1 \cap \cdots \cap V_r$ be a nonsingular complete intersection of closed hypersurfaces in $P_{n+r+1}(\mathbf{C})$ with $n \geq 3$ as in Example 2 of § 2. Then $b_2(M) = 1$, (see, for example, Hirzebruch [1, p. 161]). By Corollary 3.4, M is relatively minimal in the class of compact Kaehler manifolds which are bimeromorphic to M.

In Example 1 of § 2 we saw that if M is a compact complex manifold such that a suitable positive power K^m of its canonical line bundle K is very ample, then M is the minimal model. We prove now the following:

THEOREM 3.6.   *Let M be a compact complex manifold such that a suitable positive power K^m of its canonical line bundle K has no base points (i.e., for each point p of M, there is a holomorphic cross section of K^m which does not vanish at p). Then M is relatively minimal in its class of bimeromorphically equivalent complex manifolds.*

Proof.   Let $f : (M, N) \to (M', N')$ be a contraction, where M' is a complex manifold which is bimeromorphic to M. As in the proof of Theorem 2.3, we see that f induces a linear isomorphism from $\Gamma(K'^m)$ onto $\Gamma(K^m)$, where K' denotes the canonical line bundle of M', $\Gamma(K'^m)$ the space of holomorphic cross sections of K'^m over M', and $\Gamma(K^m)$ the space of holomorphic cross sections of K^m over M. We claim that f is nondegenerate everywhere on M. In fact, if f is degenerate at a point p of M, then $f^*(\omega')$ vanishes at p for every $\omega' \in \Gamma(K'^m)$. Since $f^* : \Gamma(K'^m) \to \Gamma(K^m)$ is an isomorphism, this means that every $\omega \in \Gamma(K^m)$ vanishes at p, in contradiction to the assumption that K^m has no base points. Since f is nondegenerate everywhere, it is a covering projection from M onto M'. On the other hand, f is bimeromorphic and hence must be biholomorphic.   QED.

Example 4.   Let M be a compact complex manifold with $c_1(M) = 0$. If $H^1(M, \Omega) = 0$ (where Ω is the sheaf of germs of holomorphic functions), in particular if M is Kaehlerian and $b_1(M) = 0$, then M is relatively minimal in its class of bimeromorphically equivalent compact complex manifolds. To prove our assertion, consider the exact sequence

$$0 \to Z \to \Omega \to \Omega^* \to 0,$$

where Ω^* is the sheaf of germs of non vanishing holomorphic functions and the homomorphism $\Omega \to \Omega^*$ is given by $\exp(2\pi i \cdot)$. Then this yields the following exact sequence:

$$\to H^1(M, \Omega) \to H^1(M, \Omega^*) \to H^2(M, Z) \to .$$

Since the homomorphism $H^1(M, \Omega^*) \to H^2(M, Z)$ maps a complex line bundle into its Chern class, a complex line bundle $\xi \in H^1(M, \Omega^*)$ with $c_1(\xi) = 0$ comes from $H^1(M, \Omega)$. If we assume that $H^1(M, \Omega) = 0$, then $c_1(\xi) = 0$ implies that ξ is a trivial bundle. Applying this result to the canonical line bundle K of M, we see that $c_1(M) = -c_1(K) = 0$ implies that K is trivial provided $H^1(M, \Omega) = 0$. If K is trivial, then M admits a nonvanishing holomorphic section. Our assertion follows from Theorem 3.6.

Example 5. Let $M = V_1 \cap \cdots \cap V_r$ be a complete intersection of nonsingular hypersurfaces $V_1, \ldots, V_r$ in $P_{n+r+1}(\mathbf{C})$ as in Example 2 of § 2. Assume $n \geq 2$ so that $b_1(M) = 0$.

If $d_1, \ldots, d_r$ are the degrees of $V_1, \ldots, V_r$ respectively and satisfy the equality $n + r + 1 = d_1 + \cdots + d_r$, then M is relatively minimal in its class of bimeromorphically equivalent compact complex manifolds (see Example 2 of § 2 and Example 4 above). In particular, a nonsingular hypersurface M of degree $n + 2$ in $P_{n+1}(\mathbf{C})$ with $n \geq 2$ is relatively minimal. Compare this with Example 2 of § 2.

IX

Miscellany

1. *Invariant Measures*

Let M be a topological space with pseudodistance d. If p is a positive real number, then the p-dimensional *Hausdorff measure* μ_p is defined as follows. For a subset E of M, we set

$$\mu_p(E) = \sup_{\varepsilon>0} \inf\left\{ \sum_{i=1}^{\infty} (\delta(E_i))^p ;\ E = \bigcup_{i=1}^{\infty} E_i,\ \delta(E_i) < \varepsilon \right\},$$

where $\delta(E)$ denotes the diameter of E. If M is a complex manifold, the invariant pseudodistances c_M and d_M defined in Chapter III induce Hausdorff measures on M. Since these pseudodistances do not increase under holomorphic mappings, the Hausdorff measures they define do not increase under holomorphic mappings. There are other invariant ways to construct intermediate dimensional measures on complex manifolds. For a systematic study of various invariant measures on complex manifolds, we refer the reader to Eisenman [1].

In this section, we shall briefly discuss invariant measures which may be considered as direct generalizations of c_M and d_M.

Let D_n be the open unit ball in $\mathbf{C}^n$. The volume element defined by the Bergman metric of D_n induces a measure on D_n, which will be denoted by μ. Theorem 4.4 of Chapter II implies that every holomorphic mapping $f : D_n \to D_n$ is measure-decreasing with respect to μ, i.e.,

$$\mu[f(E)] \leqq \mu(E) \qquad \text{for every Borel set } E \text{ in } M.$$

Let M be an n-dimensional complex manifold. Given a Borel set B in M, choose holomorphic mappings $f_i : D_n \to M$ and Borel sets E_i in D_n for $i = 1, 2, \ldots$ such that $B \subset \bigcup_i f_i(E_i)$. Then the measure $\mu_M(B)$ of B is defined by

$$\mu_M(B) = \inf \sum_i \mu(E_i),$$

where the infimum is taken over all possible choices for f_i and E_i. In analogy to c_M we define another measure γ_M as follows. For each Borel set B in M, we set

$$\gamma_M(B) = \sup_f \mu[f(B)],$$

where the supremum is taken with respect to the family of holomorphic mappings $f : M \to D_n$. From the definitions of these measures, the following proposition is evident.

PROPOSITION 1.1. *If f is a holomorphic mapping of a complex manifold M into another complex manifold N of the same dimension, then*

$$\mu_N[f(B)] \leq \mu_M(B) \quad and \quad \gamma_N[f(B)] \leq \gamma_M(B)$$

for every Borel set B in M.

COROLLARY 1.2. *If $f : M \to N$ is biholomorphic, then*

$$\mu_N[f(B)] = \mu_M(B) \quad and \quad \gamma_N[f(B)] = \gamma_M(B)$$

for every Borel set B in M.

The following two propositions follow from the fact that a holomorphic mapping of D_n into itself is measure-decreasing with respect to μ.

PROPOSITION 1.3. *For every Borel set B in a complex manifold M, we have*

$$\gamma_M(B) \leq \mu_M(B).$$

PROPOSITION 1.4. *For the unit ball D_n itself, both μ_{D_n} and γ_{D_n} coincide with the measure μ defined by the Bergman metric.*

The following proposition is trivial.

PROPOSITION 1.5.   *Let M be an n-dimensional complex manifold.*

(1) *If μ' is a measure on M such that every holomorphic mapping $f : D_n$ → M satisfies*

$$\mu'[f(E)] \leqq \mu(E) \qquad \text{for every Borel set } E \text{ in } D_n,$$

then $\mu' \leqq \mu_M$;

(2) *If μ' is a measure on M such that every holomorphic mapping $f : M$ → D_n satisfies*

$$\mu'(B) \geqq \mu[f(B)] \qquad \text{for every Borel set } B \text{ in } M,$$

then $\mu' \geqq \gamma_M$.

We call a complex manifold *M measure-hyperbolic* if $\mu_M(B)$ is positive for every nonempty open set B in M.

In contrast to Theorem 4.7 of Chapter IV, the following proposition is easy.

PROPOSITION 1.6.   *Let M be a complex manifold and $\tilde{M}$ a covering manifold of M. Then $\tilde{M}$ is measure-hyperbolic if and only if M is.*

If $\tilde{M}$ is a bounded domain in $\mathbf{C}^n$, then $\gamma_{\tilde{M}}(B)$ is positive whenever B is a nonempty open set in $\tilde{M}$. From Propositions 1.3 and 1.6, we obtain

COROLLARY 1.7.   *Let M be a complex manifold which has a bounded domain of $\mathbf{C}^n$ as a covering manifold. Then M is measure-hyperbolic.*

In analogy to Theorem 4.11 of Chapter IV, we have

THEOREM 1.8.   *Let M be an n-dimensional Hermitian manifold with Hermitian metric $ds_M{}^2 = 2 \sum g_{ij}\, dw^i\, d\bar{w}^j$ and Ricci tensor R_{ij} such that $(R_{ij}) \leqq -(cg_{ij})$ (i.e., the Hermitian matrix $(R_{ij} + cg_{ij})$ is negative semi-definite) for some positive constant c. Then M is measure-hyperbolic.*

Proof. By normalizing the metric, we may assume that $c = 1$. If $ds^2 = 2 \sum h_{ij}\, dz^i\, d\bar{z}^j$ denotes the Bergman metric in D_n, its Ricci tensor is given by $-h_{ij}$. By Theorem 4.4 of Chapter II, every holomorphic mapping $f : D_n \to M$ is volume-decreasing with respect to the volumes defined ds^2 and $ds_M{}^2$. If we denote by μ' the measure on M defined by $ds_M{}^2$, then Proposition 1.5 implies that $\mu'(B) \leqq \mu_M(B)$ for every Borel set B in M. Our assertion is now clear.   QED.

In particular, every compact Hermitian manifold with negative definite Ricci tensor is measure-hyperbolic.

THEOREM 1.9. *A compact complex manifold with negative first Chern class is measure-hyperbolic.*

Proof. Since such a manifold admits a volume element whose Ricci tensor is negative definite, the theorem follows from Lemma 4.1 of Chapter II.   QED.

Example 1. Let $M = V_1 \cap \cdots \cap V_r$ be a complete intersection of nonsingular hypersurfaces $V_1, \ldots, V_r$ of degrees $d_1, \ldots, d_r$ in $P_{n+r+1}(\mathbf{C})$ such that $d_1 + \cdots + d_r \geqq n + r + 2$. Then M is measure-hyperbolic (see Example 2 in § 2 of Chapter VIII).

THEOREM 1.10. *If a complex manifold M is hyperbolic, it is also measure-hyperbolic.*

Proof. Let μ_{2n} denote the $2n$-dimensional Hausdorff measure defined by the distance d_M. From Proposition 1.5 we obtain

$$\mu_M(B) \geqq \mu_{2n}(B).$$

On the other hand, $\mu_{2n}(B)$ is positive for every nonempty open set B (see Hurewicz and Wallman [1, Chapter VII]).   QED.

Example 2. Let M' be a measure-hyperbolic complex manifold and let $g : (M', N') \to (M, N)$ be a monoidal transformation (see § 2 of Chapter VIII). Since the contraction $f : (M, N) \to (M', N')$ is measure-decreasing and, for every nonempty open subset U of M, $f(U)$ contains a nonempty open subset of M', it follows that M is also measure-hyperbolic. This furnishes an example of a compact Kaehler manifold which is measure-hyperbolic but is not relatively minimal. It shows also that the converse to Theorem 1.9 does not hold (see Example 1 in § 2 of Chapter VIII).

Example 3. In connection with Theorem 1.9 and Example 2, we mention that an algebraic manifold of general type is measure hyperbolic (see Kobayashi and Ochiai [2]).

2. *Intermediate Dimensional-Invariant Measures*

Let M be a complex manifold of dimension n. Let B be a real k-dimensional differentiable manifold (with or without boundary) together with a differentiable mapping $\varphi : B \to M$. For each positive integer m, we define a k-dimensional measure $\mu(B, \varphi)_m$ of (B, φ) as follows. Choose a countable open cover $\{B_i\}$ of B, a differentiable mapping $h_i : B_i \to D_m$ (the open unit ball in $\mathbf{C}^m$), and a holomorphic mapping $f_i : D_m \to M$ for each i such that

$$f_i \circ h_i = \varphi\,|_{B_i}.$$

We denote by ds^2 the Bergman metric of D_m. Then $h_i{}^* ds^2$ is a positive semidefinite quadratic differential form on B_i and hence induces a (possibly degenerate) volume element, i.e., a nonnegative k-form, on B_i, which will be denoted by v_i. (The construction of the volume element from a Riemannian metric can be applied to $h_i{}^* ds^2$ even when $h_i{}^* ds^2$ is degenerate. At the points where $h_i{}^* ds^2$ is degenerate, the volume element v_i vanishes.) We set

$$\mu(B, \varphi)_m = \inf \sum_i \int_{B_i} v_i,$$

where the infimum is taken with respect to all possible choices for $\{B_i, h_i, f_i\}$. The measure $\mu(B, \varphi)_m$ can be infinite. If we cannot find $\{B_i, h_i, f_i\}$ satisfying the condition $f_i \circ h_i = \varphi\,|_{B_i}$, then we set $\mu(B, \varphi)_m = \infty$. We have clearly

$$\mu(B, \varphi)_1 \geqq \mu(B, \varphi)_2 \geqq \mu(B, \varphi)_3 \geqq \cdots.$$

If $\varphi(B)$ is contained in a compact subset of M and if $m \geqq n$, then $\mu(B, \varphi)_m$ is finite.

In analogy to the measure γ_M defined in § 1, we define another k-dimensional measure $\gamma(M, \varphi)_m$ as follows. Choose a holomorphic mapping $f : M \to D_m$. Let v_f be the volume element defined by $\varphi^*(f^* ds^2)$; it is a k-form vanishing at the points where $\varphi^*(f^* ds^2)$ is degenerate. We set

$$\gamma(B, \varphi)_m = \sup_f \int_B v_f,$$

where the supremum is taken with respect to all possible holomorphic mappings $f : M \to D_m$. Clearly, we have

$$\gamma(B, \varphi)_1 \leqq \gamma(B, \varphi)_2 \leqq \gamma(B, \varphi)_3 \leqq \cdots.$$

We shall see (Proposition 2.3) that $\gamma(B, \varphi)_m$ is finite if $\varphi(B)$ is contained in a compact subset of M.

If B is a real submanifold of M and φ is the injection of B into M, then we write $\mu(B)_m$ and $\gamma(B)_m$ for $\mu(B, \varphi)_m$ and $\gamma(B, \varphi)_m$, respectively. The following proposition is evident.

PROPOSITION 2.1. *If f is a holomorphic mapping of a complex manifold M into another complex manifold N and if B is a real k-dimensional manifold with a mapping $\varphi : B \to M$, then*

$$\mu(B, f \circ \varphi)_m \leqq \mu(B, \varphi)_m \qquad and \qquad \gamma(B, f \circ \varphi)_m \leqq \gamma(B, \varphi)_m.$$

COROLLARY 2.2. *If $f : M \to N$ is biholomorphic, then*

$$\mu(B, f \circ \varphi)_m = \mu(B, \varphi)_m \qquad and \qquad \gamma(B, f \circ \varphi)_m = \gamma(B, \varphi)_m.$$

PROPOSITION 2.3. *If M is a complex manifold and B is a real manifold with a mapping $\varphi : B \to M$, then*

$$\gamma(B, \varphi)_1 \leqq \gamma(B, \varphi)_2 \leqq \cdots \leqq \mu(B, \varphi)_2 \leqq \mu(B, \varphi)_1.$$

Proof. It suffices to prove the following inequality for every pair of positive integers p and q:

$$\gamma(B, \varphi)_p \leqq \mu(B, \varphi)_q.$$

If we make use of the notations $\{B_i, h_i, f_i\}$ and f used in the definitions of $\mu(B, \varphi)_q$ and $\gamma(B, \varphi)_p$ respectively, then we have the following commutative diagram:

$$B_i \xrightarrow{\varphi} M \xrightarrow{f} D_p.$$
$$h_i \searrow \quad \nearrow f_i$$
$$D_q$$

Denote the metrics on D_p and D_q by $ds_p{}^2$ and $ds_q{}^2$, respectively. Then

$$\varphi^*(f^* \, ds_p{}^2) = h_i{}^* f_i{}^* (f^* \, ds_p{}^2) = h_i{}^* (f_i{}^* f^* \, ds_p{}^2) \leqq h_i{}^* (ds_q{}^2),$$

where the last inequality follows from the Schwarz lemma for $f \circ f_i : D_q \to D_p$ (see Corollary 4.2 of Chapter III). On the other hand, the volume elements v_i and v_f are induced, respectively, from $h_i{}^*(ds_q{}^2)$ and $\varphi^*(f^* \, ds_p{}^2)$. Hence, $v_f \leqq v_i$ on B_i so that

$$\int_B v_f \leqq \sum_i \int_{B_i} v_i.$$

$$\text{QED.}$$

The following proposition follows also from the Schwarz lemma for holomorphic mappings $D_m \to D_n$ and $D_n \to D_m$.

PROPOSITION 2.4.   *Let B be a real manifold with a mapping* $\varphi : B \to D_n$. *Then, for* $m \geq n$, *both* $\mu(B, \varphi)_m$ *and* $\gamma(B, \varphi)_m$ *coincide with the integral over B of the volume element defined by* $\varphi^* ds^2$, *where* ds^2 *denotes the Bergman metric of* D_n.

So far, we have been using the expression "k-dimensional measure" rather loosely. If, for each real k-dimensional manifold B with a mapping $\varphi : B \to M$, $\nu(B, \varphi)$ is a nonnegative real number (including the infinity) and if $\nu(B, \varphi) = \inf \sum_i \nu(B_i, \varphi)$, where the infimum is taken with respect to all countable open covers $\{B_i\}$ of B, then we shall call ν a k-*dimensional measure* on M.

PROPOSITION 2.5.   *Let* ν *be a k-dimensional measure on a complex manifold M of dimension n. Assume* $m \geq n$.

(1) *If* $\nu(B, f \circ \psi) \leq \mu(B, \psi)_m$ *for every real k-dimensional manifold B with* $\psi : B \to D_n$ *and for every holomorphic mapping* $f : D_n \to M$, *then* $\nu(B, \varphi) \leq \mu(B, \varphi)_m$ *for every real k-dimensional manifold B with* $\varphi : B \to M$;

(2) *If* $\nu(B, \varphi) \geq \gamma(B, f \circ \varphi)_m$ *for every real k-dimensional manifold B with* $\varphi : B \to M$ *and for every holomorphic mapping* $f : M \to D_n$, *then* $\nu(B, \varphi) \geq \gamma(B, \varphi)_m$ *for every real k-dimensional manifold B with* $\varphi : B \to M$.

Proof.   (1). Let $\{U_i\}$ be a countable open cover of M with biholomorphic mappings $f_i : D_n \to U_i$. We set $B_i = \varphi^{-1}(U_i)$ and $\psi_i = f_i^{-1} \circ \varphi |_{B_i}$. From the assumption on ν and Proposition 2.4 we obtain

$$\nu(B_i, \varphi) = \nu(B_i, f_i \circ \psi_i) \leq \mu(B_i, \psi_i)_m = \int_{B_i} v_i,$$

where v_i is the volume element defined by $\psi_i^* ds^2$. Hence

$$\sum_i \nu(B_i, \varphi) \leq \sum_i \int_{B_i} v_i.$$

Taking the infimums of the both sides with respect to all possible choices for $\{U_i\}$, we obtain easily the desired inequality. The proof of (2) is similar.   QED.

We shall say that a complex manifold M is (k, m)-*hyperbolic* if $\mu(B, \varphi)_m$ is different from zero for every real k-dimensional manifold B with an imbedding $\varphi : B \to M$.

The following proposition is easy to prove.

PROPOSITION 2.6. *Let M be a complex manifold and $\tilde{M}$ a covering manifold of M. Then $\tilde{M}$ is (k, m)-hyperbolic if and only if M is.*

The proof of the following corollary is similar to that of Corollary 1.7.

COROLLARY 2.7. *Let M be a complex manifold which has a bounded domain of $\mathbf{C}^n$ as a covering space. Then M is (k, m)-hyperbolic.*

More generally, we have the following theorem, the proof of which is essentially contained in the proof of Theorem 1.9.

THEOREM 2.8. *Every hyperbolic complex manifold is (k, m)-hyperbolic.*

We shall now define norms in the kth homotpy group $\pi_k(M, x_0)$ of a complex manifold M. Let α be an element of $\pi_k(M, x_0)$ and let $\varphi : S^k \to M$ be a mapping representing α. We set

$$\mu(\alpha)_m = \inf \mu(S^k, \varphi)_m , \qquad \gamma(\alpha)_m = \inf \gamma(S^k, \varphi)_m ,$$

where the infimums are taken with respect to all φ representing α. Since S^k is compact, $\mu(\alpha)_m$ is finite if m is not less than the complex dimension of M. It is easy to verify

$$\mu(\alpha + \alpha')_m \leqq \mu(\alpha)_m + \mu(\alpha')_m , \qquad \gamma(\alpha + \alpha')_m \leqq \gamma(\alpha)_m + \gamma(\alpha')_m$$

for α, $\alpha' \in \pi_k(M, x_0)$. If $k = 1$, then $\alpha + \alpha'$ should be replaced by $\alpha\alpha'$, since $\pi_1(M, x_0)$ need not be Abelian.

It is easy to see that

$$\mu(\alpha)_m \geqq \gamma(\alpha)_m \qquad \text{for} \quad \alpha \in \pi_k(M, x_0)$$

and

$$\mu(f_* \alpha)_m \leqq \mu(\alpha)_m \qquad \text{and} \qquad \gamma(f_* \alpha)_m \leqq \gamma(\alpha)_m$$

for every $\alpha \in \pi_k(M, x_0)$ and for every holomorphic mapping f of M into another complex manifold M'.

Example 1. Let $M = \{z \in \mathbf{C}^n; 0 < r < \| z \| < 1\}$. Then $\pi_{2n-1}(M, x_0) = Z$. Let α be the generator of $\pi_{2n-1}(M, x_0)$. We claim that $\gamma(\alpha)_n$ is the volume of the hypersphere $\{z \in \mathbf{C}^n; \| z \| = r\}$ with respect to the Bergman metric of the unit ball D_n provided that $n \geq 2$. Let $\varphi : S^{2n-1} \to M$ be a mapping respresenting α. Let $f : M \to D_n$ be a holomorphic mapping and v_f the volume element of S^{2n-1} defined by $\varphi^* f^* \, ds^2$, where ds^2 is the Bergman metric of D_n. Since $n \geq 2$, f can be extended to a holomorphic mapping of D_n into D_n. Since the extended mapping $f : D_n \to D_n$ is distance-decreasing by the Schwarz lemma, it follows that $\varphi^* f^* \, ds^2 \leq \varphi^* \, ds^2$. If we denote by v the volume element of S^{2n-1} defined by $\varphi^* \, ds^2$, then

$$\gamma(S^{2n-1}, \varphi)_n = \sup_f \int_{S^{2n-1}} v_f = \int_{S^{2n-1}} v.$$

Let $S^{2n-1}(r)$ denote the sphere $\{z \in \mathbf{C}^n; \| z \| = r\}$. Considering the mapping $x \in S^{2n-1} \to r \, \varphi(x)/\| \varphi(x) \| \in S^{2n-1}(r)$, we see easily that the integral $\int_{S^{2n-1}} v$ is greater than the volume of $S^{2n-1}(r)$. On the other hand, we can find φ such that this integral is arbitrarily close to the volume of $S^{2n-1}(r)$. This completes the proof of our claim. This example shows that $\gamma(\alpha)_m$ is nontrivial sometimes.

Similarly, we can define norms (or rather, pseudonorms) in the homology groups of a complex manifold. Both $\mu(B, \varphi)_m$ and $\gamma(B, \varphi)_m$ can be defined when B is piecewise differentiable, e.g., a simplex. We can therefore define $\mu(s)_m$ and $\gamma(s)_m$ when s is a differentiable singular simplex of M. If $c = \sum_i n_i s_i$ is a differentiable singular chain, we define

$$\mu(c)_m = \sum_i | n_i | \, \mu(s_i)_m, \qquad \gamma(c)_m = \sum_i | n_i | \, \gamma(s_i)_m.$$

If $\alpha \in H_k(M, Z)$, then we define

$$\mu(\alpha)_m = \inf_c \mu(c)_m, \qquad \gamma(\alpha)_m = \inf_c \gamma(c)_m,$$

where the infimums are taken with respect to all cycles c representing the homology class α.

If B is a complex manifold and $\varphi : B \to M$ is a holomorphic mapping in the definition of $\mu(B, \varphi)_m$ given at the beginning of this section, it is natural to modify the definition of $\mu(B, \varphi)_m$ by considering only holomorphic h_i. We obtain a slightly different measure $\mu(B, \varphi)_m'$ in this way. If B is a complex space, we can still define $\mu(B, \varphi)_m'$ by ignoring the singular locus of B. The singular locus has a lower dimension anyway. In particular, if V is a complex subspace of M, then we can define $\mu(V)_m'$. If B is of complex dimension r and if $\varphi(B)$ is contained in a compact subset of M, then $\mu(B, \varphi)_m'$ is finite for $m \geq r$.

Assume that M is a Hermitian manifold with metric ds_M^2 whose holomorphic sectional curvature is bounded above by a negative constant. We shall normalize ds_M^2 in such a way that every holomorphic mapping $f : D_m \to M$ is distance-decreasing, i.e., $f^* ds_M^2 \leq ds^2$. This is possible by Theorem 4.1 of Chapter III. In defining $\mu(B, \varphi)_m'$, we choose a countable open cover $\{B_i\}$ of B and holomorphic mappings $h_i : B_i \to D_m$ and $f_i : D_m \to M$ such that $f_i \circ h_i = \varphi |_{B_i}$. Then

$$\varphi^* ds_M^2 = h_i^*(f_i^* ds_M^2) \leq h_i^* ds^2 \qquad \text{on} \quad B_i.$$

It follows that the volume element v_i on B_i constructed from $h_i^* ds^2$ is bounded below by the volume element obtained from $\varphi^* ds_M^2$. This implies that $\mu(B, \varphi)_m'$ is bounded below by the volume of B with respect to the volume element constructed from a positive semidefinite form $\varphi^* ds_M^2$. In particular, if V is a complex subspace of M, then $\mu(V)_m'$ is bounded below by the volume of V with respect to the metric induced by ds_M^2. Hence,

PROPOSITION 2.9. *Let M be a Hermitian manifold with holomorphic sectional curvature bounded above by a negative constant. Then $\mu(V)_m' > 0$ for every complex subspace V of positive dimension.*

Let M be a compact complex manifold. Every r-dimensional closed complex subspace V of M is a $2r$-dimensional cycle. A cycle of the form $c = \sum n_i V_i$, where each V_i is a closed complex subspace of dimension r, is called an *analytic r-cycle*. We say that $\sum n_i V_i$ is *positive* if all n_i's are positive. By considering only those elements of $H_{2r}(M; Z)$ which can be represented by analytic r-cycles, we obtain a subgroup $H_r'(M; Z)$ of $H_{2r}(M; Z)$. For $\alpha \in H_r'(M; \mathbf{Z})$, we set

$$\mu(\alpha)_m' = \inf_c (\sum |n_i| \mu(V_i)_m'), \qquad \gamma(a)_m' = \inf_c [\sum |n_i| \gamma(V_i)_m],$$

where the infimums are taken with respect to all analytic r-cycles $c = \sum n_i V_i$ representing α.

THEOREM 2.10. *Let M be a compact Kaehler manifold with negative holomorphic sectional curvature. If $\alpha \in H_r'(M; Z)$ can be represented by a positive analytic r-cycle, then $\mu(\alpha)_m'$ is positive.*

Proof. Let $c = \sum n_i V_i$ be any analytic cycle representing α. Let ω be the Kaehler 2-form of M. The volume of V_i with respect to ds_M^2 is given by the integral $\int_{V_i} \omega^r$. This integral does not exceed $\mu(V_i)_m'$ (if ds_M^2 is normalized as in the proof of Proposition 2.9). Hence,

$$\mu(c)_m' = \sum |n_i| \mu(V_i)_m' \geq \sum |n_i| \int_{V_i} \omega^r \geq \sum n_i \int_{V_i} \omega^r = \int_c \omega^r.$$

Since ω^r is closed, the integral $\int_c \omega^r$ depends only on α and not on c. Since α can be represented by a positive analytic cycle, this integral is positive. It follows that $\mu(c)_m'$ is bounded below by a positive constant which depends only on α.   QED.

It seems difficult to find criteria for $\mu(\alpha)_m$ to be positive. It would be of interest to investigate relationships between the pseudonorms defined above and the pseudonorm defined by Chern, Levine, and Nirenberg (see Nirenberg [1], Chern, Levine, and Nirenberg [1]). I suspect that their pseudonorm for $H_k(M; Z)$ lies between the two pseudonorms I introduced here and is close to $\gamma(\alpha)_m$ (see § 3).

3. *Unsolved Problems*

In order to state some of the unsolved problems on hyperbolic manifolds, we introduce notions closely related to that of hyperbolic manifold.

Let M and N be complex spaces and denote by $\mathrm{Hol}(N, M)$ the set of holomorphic mappings from N into M. A sequence $f_i \in \mathrm{Hol}(N, M)$ is said to be *compactly divergent* if given any compact K in N and compact K' in M, there exists j such that $f_i(K) \cap K' = \emptyset$ for all $i \geq j$. Fix a metric ϱ on M which induces its topology. $\mathrm{Hol}(N, M)$ is said to be *normal* if every sequence in $\mathrm{Hol}(N, M)$ contains a subsequence which is either uniformly convergent on compact sets or compactly divergent. According to Wu [2], M is said to be *taut* if $\mathrm{Hol}(N, M)$ is normal for every N. If $\mathrm{Hol}(N, M)$ is equicontinuous for every N with respect to some metric ϱ on M, then M is said to be *tight*.

We shall say that a complex space M is *Carathéodory-hyperbolic* or

C-hyperbolic for short (respectively, *complete C-hyperbolic*) if there is a covering space $\tilde{M}$ of M whose Carathéodory pseudodistance $c_{\tilde{M}}$ is a distance (complete distance).

We say that a complex space M admits *no complex line* if there is no holomorphic mapping from C into M other than the constant maps.

These various concepts are related in the following manner:

complete C-hyperbolic $\overset{1}{\rightleftarrows}$ C-hyperbolic

$2\downarrow$ $\qquad\qquad\qquad$ $2'\Updownarrow$

complete hyperbolic $\overset{1'}{\rightleftarrows}$ hyperbolic $\overset{4}{\rightleftarrows}$ no complex line

$3\downarrow$ $\qquad\qquad$ $3'\updownarrow$

taut $\overset{1''}{\rightleftarrows}$ tight

The implications (3), (3'), and (1'') have been proved by Kiernan [2] and Eisenman [2] independently. The other implications are either trivial or have been proved in Chapter IV. A bounded domain which is not a domain of holomorphy provides an example to show that the converses to (1), (1'), and (1'') are not true. D. Eisenman and L. Taylor have shown that the converse to (4) is not true by the following example. Let

$$M = \{(2, w) \in \mathbf{C}^2;\ |z| < 1,\ |zw| < 1\} - \{(0, w);\ |w| \geq 1\}.$$

The mapping $h : (z, w) \to (z, zw)$ maps M into the unit bidisk, and is one-to-one except on the set $z = 0$. Let $f : \mathbf{C} \to M$ be a holomorphic mapping. Then $h \circ f$ is a constant map. It follows that either f is a constant map or f maps $\mathbf{C}$ into the set $\{(0, w) \in M\}$. But this set is equivalent to $\{w \in \mathbf{C};\ |w| < 1\}$ which is hyperbolic. Hence f is a constant map in either case. To see that M is not hyperbolic, let $p = (0, b) \in M$ with $b \neq 0$. We shall show that the pseudodistance $d_M(0, p)$ between the origin 0 and p is zero. Set $p_n = [(1/n), b]$. Then $d_M(0, p) = \lim d_M(0, p_n)$. Let $a_n = \min[n, (n/|b|)^{1/2}]$. Then the mapping $t \in D \to (a_n t/n, a_n bt)M$ maps $1/a_n$ into p_n. This shows that $\lim d_M(0, p_n) \leq \lim d_D(0, 1/a_n) = 0$.

The following example by Kaup shows that the converse to (2') is not true. Consider a principal bundle $\mathbf{C}^2 - \{0\}$ over $P_1(\mathbf{C})$ and let L be the associated line bundle. L may be also obtained from $\mathbf{C}^2$ by blowing up the origin. Let B be the set of elements of L of length less than 1 with respect to the natural Hermitian fiber metric. B may be obtained from the unit ball in $\mathbf{C}^2$ by blowing up the origin. It is a unit disk bundle over $P_1(\mathbf{C})$. Choose three points in $P_1(\mathbf{C})$ and remove the closed disk of radius $\frac{1}{2}$ from the fibers over these three points. Then the resulting space is a simply connected hyperbolic manifold which is not C-hyperbolic.

It would be natural to ask the following questions.

Problem 1. Is every taut manifold complete hyperbolic? (I believe that the answer is probably affirmative.) Is every complete hyperbolic manifold complete C-hyperbolic?

For each positive integer n, let

$$
\varrho(n) = \begin{cases} \left(\dfrac{n}{2} + 1\right)^2 + 1 & \text{if } n \text{ is even,} \\[2ex] \left(\dfrac{n+1}{2}\right)\left(\dfrac{n+3}{2}\right) + 1 & \text{if } n \text{ is odd.} \end{cases}
$$

Let M be the manifold obtained from $P_n(\mathbf{C})$ by deletion of $\varrho(n)$ hyperplanes in general position. Wu [3] proved that M admits no complex line. The first few values of ϱ are: $\varrho(2) = 5$, $\varrho(3) = 7$, $\varrho(4) = 10$, $\varrho(5) = 13$. On the other hand, Kiernan [1] proved that if M is the manifold obtained from $P_n(\mathbf{C})$ by deletion of only $2n$ hyperplanes in general position, then M admits a complex line and hence is not hyperbolic. Hence, in Wu's result, $\varrho(n)$ is the best possible for $n = 2, 3$. But it may not be the best possible for $n \geq 4$.

Problem 2. Delete from $P_n(\mathbf{C})$ $\varrho(n)$ hyperplanes in general position. Is the resulting manifold M hyperbolic? Is it even C-hyperbolic? Is it possible to extend every holomorphic mapping from the punctured disk D^* into M to a holomorphic mapping from D into $P_n(\mathbf{C})$? (For this question, see Theorem 6.1 of Chapter VI.) Is it also possible to lower $\varrho(n)$ to $2n + 1$?

Let M be $\mathbf{C}^n$ minus the hypersurface $(z_1)^d + \cdots + (z_n)^d + 1 = 0$. As Kiernan [1] pointed out, the holomorphic mapping $f : \mathbf{C} \to M$ defined by $f(t) = (t, (-1)^{1/d} t, 0, \ldots, 0)$ is nonconstant. This implies that, if $z_0, z_1, \ldots, z_n$ denotes the homogeneous coordinate system for $P_n(\mathbf{C})$, then $P_n(\mathbf{C})$ minus the hypersurface $(z_1)^d + \cdots + (z_n)^d = 0$ admits a complex line for any degree d. But this hypersurface is rather special.

Problem 3.   Let M be $P_n(\mathbf{C})$ minus a "generic" hypersurface of high degree d. Is M hyperbolic? We may further ask questions similar to those in Problem 2.

Problem 4.   Is a generic hypersurface of large degree in $P_n(\mathbf{C})$ hyperbolic? Is it at least without complex line? The surface $(z_0)^d + (z_1)^d = (z_2)^d + (z_3)^d$ contains a rational curve

$$z_0 = z_2 = u, \qquad z_1 = z_3 = v.$$

Problem 5.   Does a (complete) hyperbolic manifold M admit a (complete) Hermitian metric whose holomorphic sectional curvature is bounded above by a negative constant? (This is the converse of Theorem 4.11 of Chapter IV.) A related question is the following. If the universal covering manifold $\tilde{M}$ of a complex manifold M admits a (complete) Hermitian metric whose holomorphic sectional curvature is bounded above by a negative constant, does M itself admit such a Hermitian metric? Because of Theorem 4.7 of Chapter IV, if the answer to the last question is negative, the answer to the first question will be also negative.

Let X be a compact complex space and M a complex space admitting an analytic subset of a bounded domain in $\mathbf{C}^n$ as a covering space. Let $a \in X$. If f and g are holomorphic mappings from X into M such that $f(a) = g(a)$ and $f_*(u) = g_*(u)$ for all $u \in \pi_1(X, a)$, then $f = g$ (Borel and Narasimhan [1]). Actually, Borel and Narasimhan proved a little more.

Problem 6.   It is probably possible to extend the result of Borel and Narasimhan to the case where M is C-hyperbolic. Will it be also possible to extend it to the case where M is hyperbolic?

Problem 7.   Is a compact C-hyperbolic or hyperbolic manifold projective-algebraic?

In connection with the problem above, we pose

Problem 8.   Classify the two-dimensional compact hyperbolic manifolds.

Let (Y, p, X) be a complex analytic family of complex structures on a compact manifold. In other words, X and Y are complex manifolds and $p : Y \to X$ is a surjective proper holomorphic mapping whose differential p_* is of maximal rank ($=\dim X$) everywhere. Denote a fiber

$p^{-1}(x)$, $x \in X$, by M_x. The complex manifolds M_x, $x \in X$, are all compact and diffeomorphic to each other, but not biholomorphic to each
other in general.

Problem 9.   If all fibers M_x except one fiber M_o are hyperbolic, is M_o
also hyperbolic?

Problem 10.   If one fiber M_o is hyperbolic, does there exist an open
neighborhood U of o in X such that $p^{-1}(U)$ is (complete) hyperbolic?
[If $p^{-1}(U)$ is hyperbolic and V is a smaller neighborhood which is complete hyperbolic, then $p^{-1}(V)$ is complete hyperbolic.]
In connection with Problem 10, we can ask also the following question.

Problem 11.   Let M be a compact manifold with two complex structures J and J'. If (M, J) is hyperbolic, is (M, J') also hyperbolic?

Problem 12.   Investigate the structures of a homogeneous hyperbolic
manifold. Is there any other than the homogeneous bounded domains
in $\mathbf{C}^n$? If $M = G/H$ is a homogeneous hyperbolic manifold, then M is
noncompact by Theorem 2.1' of Chapter V and G has no nondiscrete
center by Theorem 2.2 of Chapter V. (If a holomorphic vector field Z
on M generates a one-parameter group belonging to the center of G,
then iZ will also generate a global one-parameter group of transformations of M in contradiction to Theorem 2.2 of Chapter V unless $Z = 0$.)

Problem 13.   Extend Theorem 6.2 of Chapter VI to the case where A
is an analytic subset with singular points.
We shall now elaborate on the statement made concerning the intrinsic
seminorms of Chern, Levine, and Nirenberg in § 2. Let u be a real C^2
function on a complex manifold M. We define a real k-form, $1 \leq k \leq 2n$
$= 2 \dim M$, $\omega^k(u)$ as follows:

$$\omega^{2p-1}(u) = d^c u \wedge (dd^c u)^{p-1},$$
$$\omega^{2p}(u) = du \wedge d^c u \wedge (dd^c u)^{p-1},$$

where $d^c = i(\bar{\partial} - \partial)$. We denote by $F_k = F_k(M)$ the set of real C^2
functions u satisfying the following three conditions:

(i) $0 < u < 1$;
(ii) u is plurisubharmonic, i.e., $(\partial^2 u / \partial z^j \, \partial \bar{z}^k)$ is positive semidefinite;
(iii) $d[\omega^k(u)] = 0$.

Let B be a real k-dimensional differentiable manifold and $\varphi : B \to M$ a differentiable mapping. We define $v_k(B, \varphi)$ by

$$v_k(B, \varphi) = \sup_{u \in F_k} \int_B \varphi^*[\omega^k(u)].$$

Then v_k is a k-dimensional measure on M in the sense defined in § 2. If f is a holomorphic mapping from M into another complex manifold, then

$$v_k(B, f \circ \varphi) \leqq v_k(B, \varphi).$$

  Problem 14.   Does there exist a (universal) constant c so that

$$\gamma(B, \varphi)_n \leqq c \, v_k(B, \varphi) \leqq \mu(B, \varphi)_n$$

for k-dimensional B and mappings $\varphi : B \to M$? In view of Proposition 2.5 it suffices to prove these inequalities in the special case where M is the open unit ball D_n in $\mathbf{C}^n$. To prove the second inequality, it would be necessary to obtain an estimate similar to the one proved by Chern, Levine, and Nirenberg [1]. Their basic inequality says that if K is an open subset with compact closure in D_n, then

$$\int_K (|\, u_j\,|^2 + |\, \text{any minor of } \{u_{j\bar{k}}\}\,|) \, dV \leq C$$

with C a fixed constant independent of u. Here $u_j = \partial u / \partial z^j$, $u_{j\bar{k}} = \partial^2 u / \partial z^j \, \partial \bar{z}^k$, and dV represents the usual volume element in $\mathbf{C}^n$.

Bibliography

Accola, R. D. M.
[1] Differentials and extremal length on Riemann surfaces, *Proc. Natl. Acad. Sci.*, **46** (1960), 540–543.
[2] Automorphisms of Riemann surfaces, *J. Anal. Math.*, **18** (1967), 1–5.
[3] On the number of automorphisms of a closed Riemann surface, *Trans. Am. Math. Soc.*, **131** (1968), 398–408.

Ahlfors, L. V.
[1] An extension of Schwarz's lemma, *Trans. Am. Math. Soc.*, **43** (1938), 359–364.
[2] *Complex Analysis*, McGraw-Hill, New York, 1953.

Andreotti, A.
[1] Sopra il problema dell'uniformizzazione per alcune classi di superficie algebriche, *Rend. Accad. Nazl. Pisa*, **XI** (4) 2 (1941), 111–127.

Andreotti, A., and Frankel, T. T.
[1] The Lefschetz theorem on hyperplane sections, *Ann. Math.*, **69** (1959), 713–717.

Andreotti, A., and Stoll, W.
[1] Extension of holomorphic maps, *Ann. Math.*, **72** (1960), 312–349.

Baily, W. L., Jr.
[1] The decomposition theorem for V-manifolds, *Am. J. Math.*, **78** (1956), 862–888.
[2] On the imbedding of V-manifolds in projective space, *Am. J. Math.*, **79** (1957), 403–430.
[3] On Satake's compactification of V_n, *Am. J. Math.*, **80** (1958), 348–364.

Baily, W. L., Jr., and Borel, A.
[1] Compactification of arithmetic quotients of bounded symmetric domains, *Ann. Math.*, **84** (1966), 442–528.

Bergman, S.
[1] Uber die Kernfunktion eines Bereiches und ihr Verhalten am Rande, *J. Reine Angew. Math.*, **169** (1933), 1–42; **172** (1935), 89–128.

Bishop, R. L., and Goldberg, S. I.
[1] On the second cohomology group of a Kaehler manifold of positive curvature, *Proc. Am. Math. Soc.*, **16** (1965), 119–122.

Bochner, S.
[1] On compact complex manifolds, *J. Indian Math. Soc.*, **11** (1947), 1–21.

Bochner, S., and Martin, W. T.
[1] *Several Complex Variables*, Princeton Univ. Press, Princeton, N.J., 1948.

Bochner, S., and Montgomery, D.
   [1] Locally compact groups of differentiable transformations, *Ann. Math.*, **47** (1946), 639–653.
   [2] Groups on analytic manifolds, *Ann. Math.*, **48**, 659 (1947).
Borel, A., and Narasimhan, R.
   [1] Uniqueness conditions for certain holomorphic mappings, *Inventiones Math.*, **2** (1967), 247–255.
Bott, R., and Chern, S. S.
   [1] Hermitian vector bundles and the equidistribution of the zeros of their holomorphic sections, *Acta Math.*, **114** (1965), 71–112.
Bremermann, H. J.
   [1] Holomorphic continuation of the kernel function and the Bergman metric in several complex variables, in *Lectures on Functions of a Complex Variable*, Univ. of Michigan Press, Ann Arbor, 1955, pp. 349–383.
   [2] Die Characterisierung von Regularitätsgebieten durch pseudokonvexe Funktionen, *Schriftenreihe Math. Inst. Univ. Münster*, No. 5 (1951), pp. 1–92.
Carathéodory, C.
   [1] Über das Schwarzsche Lemma bei analytischen Funktionen von zwei komplexen Veränderlichen, *Math. Ann.*, **97** (1926), 76–98.
   [2] Über die Geometrie der analytischen Abbildungen, die durch analytische Funktionen von zwei Veränderlichen vermittelt werden, *Abhandl. Math. Sem. Univ. Hamburg*, **6** (1928), 97–145.
   [3] Über die Abbildungen, die durch Systeme von analytische Funktionen von mehreren Veränderlichen erzeugt werden, *Math. Z.*, **34** (1932), 758–792.
Cartan, E.
   [1] Sur les domaines bornés de l'espace de *n* variables complexes, *Abhandl. Math. Sem. Univ. Hamburg*, **11** (1935), 116–162.
Cartan, H.
   [1] Sur les groupes de transformations analytiques, *Actualités Sci. Ind.*, **9**, (1935), Hermann, Paris.
   [2] Les fonctions de deux variables complexes et le problème de la représentation analytique, *J. Math. Pures appl.*, **10** (1931), 1–114.
   [3] Sur les fonctions de plusieurs variables complexes, l'itération des transformations intérieures d'un domaine borné, *Math. Z.*, **35** (1932), 760–773.
   [4] Quotient d'un espace analytique par un groupe d'automorphismes, *Algebraic Geometry and Topology, Symp. in Honor of Lefschetz*, Princeton Univ. Press, Princeton, N.J., 1957, pp. 90–102.
   [5] Quotients of complex analytic spaces, in *Contributions to Function Theory*, Tata Institute, Bombay, 1960, 1–15.
Chern, S. S.
   [1] Characteristic classes of Hermitian manifolds, *Ann. Math.*, **47** (1946), 85–121.
   [2] On holomorphic mappings of Hermitian manifolds of the same dimension, *Proc. Symp. Pure Math., Vol.* 11 (1968), *Entire Functions and Related Parts Analysis*, pp. 157–170, Amer. Math. Soc., Providence, R. I.

[3] The integrated form of the first main theorem for complex analytic mappings in several complex variables, *Ann. Math.*, **71** (1960), 536–551.

[4] Complex analytic mappings of Riemann surfaces I, *Am. J. Math.*, **82** (1960), 323–337.

[5] Holomorphic mappings of complex manifolds, *Enseigment Math.*, **7** (1961), 179–187.

Chern, S. S., Levine, H. I., and Nirenberg, L.

[1] Intrinsic norms on a complex manifold, in *Collected Papers in Honor of K. Kodaira*, Springer, Berlin, 1969.

Dinghas, A.

[1] Ein *n*-dimensionales Analogon des Schwarz–Pickschen Flächensatzes für holomorphe Abbildungen der komplexen Einheitskugel in eine Kähler-Mannigfaltigkeit, *Arbeitsgemeinschaft für Forschung des Landes Nordrhein–Westfalen*, **33** (1965), 477–494.

[2] Über das Schwarzsche Lemma und verwandte Sätze, *Israel J. Math.*, **5** (1967), 157–169.

[3] Verzerrungssätze bei holomorphen Abbildungen von Hauptbereichen automorpher Gruppen mehrerer komplexer Veränderlicher in eine Kähler-Mannigfaltigkeit, *Sitzber. Heidelberg. Akad. Wiss.* (1968), 1–21.

Earle, C. J.

[1] Invariant metrics and fixed point theorems for holomorphic mappings, Summer Institute on Global Analysis, Berkeley, 1968.

Eisenman, D.

[1] Intrinsic measures on complex manifolds and holomorphic mappings, *Mem. Am. Math. Soc.*, No. 96 (1970).

[2] Holomorphic mappings into tight manifolds, *Bull. Am. Math. Soc.*, **76** (1970), 46–48.

Ford, L. R.

[1] *Automorphic Functions*, McGraw-Hill, New York, 1929.

Gindikin, S. G., Pjateckii-Sapiro, I. I., and Vinberg, E. B.

[1] Homogeneous Kaehler manifolds, in *Geometry of Homogeneous Bounded Domains*, C.I.M.E., 3° Ciclo, Urbino, Italy, 1967, pp. 3–87.

Goldberg, S. I., and Kobayashi, S.

[1] On holomorphic bisectional curvature, *J. Diff. Geometry*, **1** (1967), 225–233.

Goto, M.

[1] On algebraic homogeneous spaces, *Am. J. Math.*, **76** (1954), 811–818.

Grauert, H.

[1] Characterisierung der Holomorphiegebiete durch die vollständige Kählersche Metrik, *Math. Ann.*, **131** (1956), 38–75.

[2] Über Modifikationen und exzeptionelle analytische Mengen, *Math. Ann.*, **146** (1962), 331–368.

Grauert, H., and Reckziegel, H.

[1] Hermitesche Metriken und normale Familien holomorpher Abbildungen, *Math. Z.*, **89** (1965), 108–125.

Grauert, H., and Remmert, R.
[1] Komplexe Räume, *Math. Ann.*, **136** (1958), 245–318.

Gunning, R. C., and Rossi, H.
[1] Analytic functions of several complex variables, Prentice-Hall, Englewood Cliffs, N.J., 1965.

Hahn, K. T., and Mitchell, J.
[1] Generalization of Schwarz–Pick lemma to invariant volume in Kahler manifolds, *Trans. Am. Math. Soc.*, **128** (1967), 221–231.
[2] The same title, II, *Can. J. Math.*, **21** (1969), 669–674.

Hano, J., and Kobayashi, S.
[1] A fibering of a class of homogeneous spaces, *Trans. Am. Math. Soc.*, **94** (1960), 233–243.

Hawley, N. S.
[1] A theorem on compact complex manifolds, *Ann. Math.*, **52** (1950), 637–641.

Heins, M.
[1] *Selected Topics in the Classical Theory of Functions of a Complex Variable*, Holt, Rinehart & Winston, New York, 1962.

Helgason, S.
[1] *Differential Geometry and Symmetric Spaces*, Academic Press, New York, 1962.

Hirzebruch, F.
[1] Topological methods in algebraic geometry, Springer, Berlin, 1966.

Holmann, H.
[1] Komplexe Räume mit komplexen Transformationsgruppen, *Math. Ann.*, **150** (1963), 327–360.

Horstmann, H.
[1] Carathéodorysche Metrik und Regularitätshullen, *Math. Ann.*, **108** (1933), 208–217.

Hua, L. K.
[1] Harmonic analysis of functions of several complex variables in the classical domains, in *Translations of Mathematical Monographs*, Vol. 6, American Mathematical Society, Providence, R. I. (1963).

Huber, H.
[1] Uber analytische Abbildungen von Ringgebieten in Ringgebiete, *Compos. Math.*, **9** (1951), 161–168.
[2] Uber analytische Abbildungen Riemannscher Flächen in sich, *Comment. Math. Helv.*, **27** (1953), 1–72.

Hurewicz, W., and Wallman, H.
[1] *Dimension Theory*, Princeton Univ. Press, Princeton, N.J., 1942.

Hurwitz, A.
[1] Uber algebraische Gebilde mit eindeutigen Transformationen in sich, *Math. Ann.*, **41** (1893), 403–442.

Igusa, J.
[1] On the structure of a certain c'ass of Kähler manifolds, *Am. J. Math.*, **76** (1954), 669–673.

Jenkins, J. A.
[1] Some results related to extremal length, *Ann. Math. Studies*, No. 30 (1953), pp. 87–94.

Jordan, S.
[1] *Chern–Levine–Nirenberg's Norm on a Complex Manifold*, Thesis, Univ. of Calif., Berkeley, 1970.

Kaup, W.
[1] Reele Transformationsgruppe und invariante Metriken auf komplexen Raumen, *Investiones Math.*, **3** (1967), 43–70.
[2] *Holomorphe Abbildungen in hyperbolische Räume*, Centro Internazionale Mat. Estivo, 1967, pp. 111–123.
[3] Hyperbolische komplexe Räume, *Ann. Inst. Fourier (Grenoble)*, **18** (1968), 303–330.

Kiernan, P. J.
[1] Hyperbolic submanifolds of complex projective space, *Proc. Am. Math. Soc.*, **22** (1969), 603–606.
[2] On the relations between taut, tight and hyperbolic manifolds, *Bull. Am. Math. Soc.*, **76** (1970), 49–51.
[3] Quasiconformal mappings and Schwarz's lemma, *Trans. Am. Math. Soc.*, **147** (1970).
[4] Some remarks on hyperbolic manifolds, *Proc. Am. Math. Soc.*, **25** (1970).

Kobayashi, S.
[1] Geometry of bounded domains, *Trans. Am. Math. Soc.*, **92** (1959), 267–290.
[2] Volume elements, holomorphic mappings and the Schwarz lemma, in *Proc. Symp. Pure Math., Vol.* 11 (1968), *Entire Functions and Related Parts Analysis*, pp. 253–260. Amer. Math. Soc., Providence, R. I.
[3] Distance, holomorphic mappings and the Schwarz lemma, *J. Math. Soc. Japan*, **19** (1967), 481–485.
[4] Invariant distances on complex manifolds and holomorphic mappings, *J. Math. Soc. Japan*, **19** (1967), 460–480.
[5] On the automorphism group of a certain class of algebraic manifolds, *Tohoku Math. J.*, **11** (1959), 184–190.

Kobayashi, S., and Nomizu, K.
[1] Foundations of differential geometry, in *Interscience Tracts*, No. 15, Vol. I (1963), Vol. II (1968). Wiley, New York.

Kobayashi, S. and Ochiai, T.
[1] Satake compactification and the great Picard theorem, *J. Math. Soc. Japan*, to appear.
[2] Mappings into compact complex manifolds with negative first Chern class, *J. Math. Soc. Japan*, to appear.

Kodaira, K.
[1] On Kähler varieties of restricted type, *Ann. Math.*, **60** (1954), 28–48.

Kodaira, K., and Spencer, D. C.
[1] Groups of complex line bundles over compact Kähler varieties. Divisor class groups on algebraic varieties, *Proc. Natl. Acad. Sci. U.S.A.*, **39** (1953), 868–877.

Koranyi, A.
[1] A Schwarz lemma for bounded symmetric domains, *Proc. Am. Math. Soc.*, **17** (1966), 210–213.

Koszul, J. L.
[1] Sur la forme Hermitienne canonique des espaces homogènes complexes, *Can. J. Math.*, **7** (1955), 562–576.

Kwack, M. H.
[1] Generalization of the big Picard theorem, *Ann. Math.*, **90** (1969), 9–22.

Landau, H. J., and Osserman, R.
[1] Some distortion theorems for multivalent mappings, *Proc. Am. Math. Soc.*, **10** (1959), 87–91.
[2] On analytic mappings of Riemann surfaces, *J. Anal. Math.*, **7** (1959/60), 249–279.

Lang, S.
[1] Introduction to algebraic geometry, in *Interscience Tracts*, No. 5 (1958). Wiley, New York.

Lelong, P.
[1] Domaines convexes par rapport aux fonctions plurisousharmoniques, *J. Anal. Math.*, **2** (1952), 178–208.
[2] Fonctions plurisousharmoniques et fonctions analytiques de variables réeles, *Ann. Inst. Fourier (Grenoble)*, **11** (1961), 515–562.

Levine, H.
[1] A theorem on holomorphic mappings into complex projective space, *Ann. Math.*, **71** (1960), 529–535.

Lichnerowicz, A.
[1] Variétés complexes et tenseur de Bergmann, *Ann. Inst. Fourier (Grenoble)*, **15** (1965), 345–407.

Marden, A., Richards, I., and Rodin, B.
[1] Analytic self-mappings of Riemann surfaces, *J. Anal. Math.*, **18** (1967), 197–225.

Moisezon, B. G.
[1] On N-dimensional compact complex varieties with N algebraically independent meromorphic functions, *Am. Math. Soc. Transl.*, Ser. 2, **63** (1967), 51–177.

Narasimhan, R.
[1] Introduction to the theory of analytic spaces, *Lecture Notes in Mathematics*, No. 25 (1966), Springer, Berlin.

Nirenberg, L.
[1] Intrinsic norms on complex analytic manifolds, *Ist. Nazl. Alta Mat. Symp. Mat.*, **II**, 227 (1968).

Oka, K.
[1] Domaines pseudoconvexes, *Tohoku Math. J.*, **49** (1942), 15–52.

[2] Sur les fonctions analytiques de plusieurs variables, VII. Lemme fondamental, *J. Math. Soc. Japan*, **3** (1951), 204–214; 259–278.

Ostrowski, A.

[1] Asymptotische Abschatzung des absoluten Betrage einer Funktion, die die Werte 0 und 1 nicht annimt, *Comment. Helv. Math.*, **5** (1933), 55–87.

Peschl, E.

[1] Über den Cartan–Carathéodoryschen Eideutigkeitssatz, *Math. Ann.*, **119** (1943), 131–139.

Peters, K.

[1] Über holomorphe und meromorphe Abbildungen gewisser kompakter komplexer Mannigfaltigkeiten, *Arch. Math.*, **15** (1964), 222–231.

Pfluger, A.

[1] Über numerische Shranken im Schottky'schen Satz, *Comment. Math. Helv.*, **7** (1935), 159–170.

Poincaré, H.

[1] Sur un théorème de M. Fuchs, *Acta Math.*, **7** (1885), 1–32.

Pyatetzki-Shapiro, I. I.

[1] *Geometrie des domaines classiques et theorie des fonctions automorphes*, Dunod, Paris, 1966.

Reiffen, H. J.

[1] Die differentialgeometrischen Eigenschaften der invarianten Distanzfunktion von Carathéodory, *Schrift Math. Inst. Univ. Münster*, No. 26 (1963).

[2] Die Carathéodorysche Distanz und ihre zugehörige Differentialmetrik, *Math. Ann.*, **161** (1965), 315–324.

Remmert, R.

[1] Holomorphe und meromorphe Abbildungen komplexer Räume, *Math. Ann.*, **133** (1957), 328–370.

Royden, H. L.

[1] Report on the Teichmüller metric, *Proc. Natl. Acad. Sci. USA*, **65** (1970), 497–499.

Sampson, J. H.

[1] A note on automorphic varieties, *Proc. Natl. Acad. Sci. U.S.A.*, **38** (1952), 895–898.

Satake, I.

[1] On a generalization of the notion of manifolds, *Proc. Natl. Acad. Sci. U.S.A.*, **42** (1956), 359–363.

[2] On the compactification of the Siegel space, *J. Indian Math. Soc.*, **20** (1956), 259–281.

Schiffer, M.

[1] On the modulus of doubly connected domains, *Quart. J. Math.*, **17** (1946), 197–213.

Schottky, F.

[1] Uber den Picardschen Satz und Borelschen Ungleichungen, *Sitz. Ber. Preuss. Akad. Wiss.* (1904), 1244–1263.

Schwarz, H. A.

[1] Über diejenigen algebraischen Gleichungen zwischen zwei veränderlichen Grossen, welche ein Schar rationaler eindeutige umkehrbarer Transformationen in sich selbst zulassen, *Crelle's J.*, **87** (1879), 139–145.

Shioda, T.

[1] On algebraic varieties uniformizable by bounded domains, *Proc. Japan Acad.*, **36** (1963), 617–619.

Springer, G.

[1] *Introduction to Riemann Surfaces*, Addison Wesley, Reading, Mass., 1957.

Stein, K.

[1] Meromorphic mappings, *Enseignment Math.*, **14** (1968), 29–46.

Stoll, W.

[1] About the universal covering of the complement of a complete quadrilateral, *Proc. Am. Math. Soc.*, **22** (1969), 326–327.

[2] Über meromorphe Abbildungen komplexer Räume I & II, *Math. Ann.*, **136** (1958), 201–239; 272–316.

Tashiro, Y.

[1] The curvature of the Bergman metric, *Sci. Rept. Tokyo Kyoiku Daigaku*, **8** (1963), 280–296.

van Dantzig, D., and van der Waerden, B. L.

[1] Über metrisch homogenen Räume, *Abhandl. Math. Sem. Univ. Hamburg*, **6** (1928), 374–376.

Vinberg, E. B., and Gindikin, S. G.

[1] Kaehlerian manifolds admitting a transitive solvable automorphism group, *Mat. Sb.* (*English Transl.*), **74** (116) (1967), 333–351.

Vinberg, E. B., Gindikin, S. G., and Pyatetzki-Shapiro, I. I.

[1] Classification and canonical realization of complex bounded homogeneous domains, *Trans. Moscow Math. Soc.*, **12** (1963), 404–437.

Weil, A.

[1] Variétés abéliennes et courbes algébriques, *Actualités Sci. Ind.* (1948). Hermann, Paris.

Wu, H.

[1] Mapping of Riemann surfaces (Nevanlinna theory), *Proc. Symp. Pure Math.*, Vol. 11 (1968), *Entire Functions and Related Parts Analysis*, pp. 480–532.

[2] Normal families of holomorphic mappings, *Acta Math.*, **119** (1967), 193–233.

[3] An *n*-dimensional extension of Picard's theorem, *Bull. Am. Math. Soc.*, **75** (1969), 1357–1361.

[4] The Equidistribution Theory of Holomorphic Curves, Princeton Univ. Press, Princeton, N.J., 1970.

Zumbrunn, J. R.

[1] *On Minimal Models of Complex Varieties*, thesis, Univ. of Calif., Berkeley, 1968.

Summary of Notations

We summarize only those notations that are used most frequently throughout this monograph.

$\mathbf{C}$ The field of complex numbers.

$\mathbf{C}^n$ Vector space of n-tuples of complex numbers $(z^1, \ldots, z^n)$.

D Open unit disk in $\mathbf{C}$, i.e., $\{z \in \mathbf{C}; \ |z| < 1\}$.

D^* Punctured disk $D - \{0\}$.

D^n Polydisk $D \times \cdots \times D$ (n times).

D_n Open unit ball in C^n, i.e., $\{(z^1, \ldots, z^n); \ \Sigma \, |z^j|^2 < 1\}$.

$P_n(\mathbf{C})$ Complex projective space of dimension n.

ϱ Poincaré distance (non-Euclidean distance) in D.

d_M the pseudodistance of M defined in § 1 of Chapter IV.

c_M the Carathéodory pseudodistance of M defined in § 2 of Chapter IV.

Author Index

Numbers in italics show the page on which the complete reference is listed.

Subject Index

A

ample, 110
analytic germ, 95
analytic polyhedron, 54
  generalized, 54
analytic set, 95
  reducible, 96

B

base point, 110
Bergman kernel form, 18
Bergman kernel function, 19
Bergman metric, 18
big Picard theorem, 79
bimeromorphic, 108

C

C-hyperbolic, 130
Carathéodory hyperbolic, 129
Carathéodory pseudodistance, 49
compactly convergent, 129
complete metric space, 53
complex space, 97
complex V-manifold, 102
contraction, 108
convex hull, 55
convex with respect to F, 55
covariant vector bundle, 110

D

dimension of a complex space, 97
divisor, 112
  group of ———s, 112
generalized analytic polyhedron, 54
group of divisors, 112
group of linear equivalence, 113

H

Hausdorff measure, 119
Hermitian fiber metric, 37
Hermitian vector bundle, 37
holomorphic bisectional curvature, 39
holomorphic sectional curvature, 39
holomorphically convex, 55
hyperbolic, 57
  C-, 130
  (k, m)-, 126
  measure-, 121

K

k-dimensional measure, 125
(k, m)-hyperbolic, 126

L

liftable (mapping), 103
  locally, 103
linear equivalence, 113
  group of, 113

147